Mindfulness and Letting Be

Mindfulness and Letting Be

On Engaged Thinking and Acting

Fred Dallmayr

LEXINGTON BOOKS
Lanham • Boulder • New York • London

Published by Lexington Books
An imprint of The Rowman & Littlefield Publishing Group, Inc.
4501 Forbes Boulevard, Suite 200, Lanham, Maryland 20706
www.rowman.com

16 Carlisle Street, London W1D 3BT, United Kingdom

British Library Cataloguing in Publication Information Available

Library of Congress Cataloging-in-Publication Data

Dallmayr, Fred R. (Fred Reinhard), 1928-
Mindfulness and letting be : on engaged thinking and acting / Fred Dallmayr.
pages cm.
Includes bibliographical references and index.
ISBN 978-0-7391-9986-2 (cloth : alk. paper) -- ISBN 978-0-7391-9987-9 (electronic)
1. Philosophy, Modern--20th century. 2. Philosophy, Modern--21st century. 3. Cosmopolitanism. 4. Thought and thinking. 5. Attention. 6. Action theory. I. Title.
B804.D275 2014
128--dc23
2014026217

Printed in the United States of America

Nicht mehr für Ohren . . . : Klang,
der, wie ein tieferes Ohr,
uns, scheinbar Hörende, hört.
(No longer for ears . . . : sound
which, like a deeper ear,
hears us seemingly hearers.)
—Rainer Maria Rilke, *Gong*
(my translation)

Contents

Preface

This book is meant as a protest against the extreme noise-level of our age, the incessant chatter and cacophony invading every nook and cranny of our lives. One can hardly take a step in this world without being confronted with loudly proclaimed manifestoes, marching orders, or ideological blueprints. Any word spoken publicly immediately turns the world into a resonance chamber filled with shouts, opinion pieces, invectives, and insults.

"Mindfulness," as used in this book, is not a mind filled to the brim with opinions, doctrines, and solutions; it does not denote a predatory reason seeking to engulf everything it encounters. Rather, it means a stance of quiet abstinence, an outlook seeking to recover its bearings through reticence and "letting-be." Martin Heidegger has stipulated three things needed for mindfulness: rigor of reflection, carefulness of expression, and frugality with words. He also stated that mindful thinking is no longer philosophy in the sense of metaphysics; rather it is a thinking "on the descent into the poverty of its provisional being."

The book is my attempt to be mindful of this descent. As always, I am grateful to all my companions on this path—family, friends, and acquaintances.

Introduction

Mindfulness and Mind-Fasting

In popular discourse and the popular media, the term "mindfulness" has recently come into vogue. As employed in these contexts, the term reflects a desire to regain a degree of mental health in the midst of the stressful demands of modern life. Given the pervasiveness of stress, the desire is widespread. Thus, new institutes or learning centers have sprung up in many places offering educational relief from stress, while new journals or websites seek to keep readers abreast of developments.[1] Despite the evident need and the good intentions of many of the participants, I want to mention three dubious implications of these new initiatives. First of all, mindfulness here designates basically a care for one's own mind or mental health, irrespective of prevailing social and political conditions. This means that the accent is on individual private life, in line with the motto of "minding one's own business." This accent implies, secondly, that there is in effect a selfhood or self-being which can be cultivated and calmed given proper instruction or guidance. This assumption reflects the legacy of the Cartesian *ego cogitans*, the thinking mind separated from the world it inhabits. Governed by this legacy, mindfulness here subscribes to a whole series of binaries or antinomies, including the binaries of subject and object, selfhood and world, mind and matter. It is chiefly in the latter respect, that the term signals the aloofness of mind from nature which has troubled Western modernity.

In the present volume, I use "mindfulness" in a very different sense. Most importantly, the term does not refer to a self-contained substance or entity called "mind" which could be the target of specialized analysis. Rather, the term is used in a verbal and transitive sense, such that "mind" is seen as "minding" not so much itself but whatever it encounters. Martin Heidegger here offers guidance. His portrayal of human *Dasein* as "being-in-the-world" motivated by "care" means that *Dasein* constitutively (and not accidentally or incidentally) "minds" the world including fellow-beings and nature. The fact that *Dasein* also, and primarily, cares about "Being" does not imply a narrowly centered self-care, but rather the need to "mind" the Being-question: what it means for *Dasein* to "be." Here, minding the Being-question means to be attentive to the "ground" of Being—which again is not a fixed substance but rather an "un-ground," a creative springboard for ever new possibilities. As should be

1

noted, "mind" in the preceding senses is not an engine of acquisition, a predator ready to appropriate whatever it encounters; rather, it has to practice the difficult task of renunciation, of letting beings "*be*." This task is particularly crucial when mindfulness encounters the "un-ground" of "non-being." At this point, mind has to "un-mind" itself by emptying itself of all pretended knowledge. Mindfulness has to give way to "mind-fasting."

As presented here, mindfulness is meant not only to relieve the stress induced by the busyness of modern life (although it may do this too). Rather, its cultivation pursues a broader and deeper aim: to counteract the aimlessness of modern existence, and especially the dehumanization engendered by the near-exclusive emphasis of modernity on technical making and production. Buttressed by the Cartesian legacy of the *cogito*, modern thinking is increasingly transformed into a calculating rationality disengaged from ethical, social and broader philosophical concerns; aided and abetted by advances in technological surveillance, this rationality seeks its perfection in the attainment of total knowledge of everything (omniscience) which, in turn, makes possible total control of everything (omnipotence). Overpowered by technical rationality, genuine mindfulness is bound to wither—and with it wither caring humanity and the traditional "humanities." As a glance at the educational programs in nearly all countries around the globe confirms, developments have already reached a crisis point, a precipice from which it will be hard to retreat. As Heidegger says at one point: the nuclear Armageddon may not yet have happened, but the existence and continued production of nuclear weapons may already have eroded humanity's conscience and mindfulness.[2]

The chapters in this study approach the prospect of human mindfulness from different angles. Chapter 1 addresses the issue of mindful thinking with a focus on the status of philosophy and philosophical "theory." The chapter examines modern rational theorizing against the backdrop of premodern—classical and medieval—philosophy (in the West) when thinkers were still participants, and not detached onlookers or spectators in the world. This older tradition came to a halt with the rise of modern science whose outlook was refined in Enlightenment rationalism and educationally stabilized in nineteenth-century scientism and positivism. With these developments, humanistic mindfulness was increasingly placed on the defensive. The chapter at this point discusses a number of initiatives seeking to vindicate and re-invigorate the role of mindful thinking in the twentieth century: Maurice Merleau-Ponty's *Éloge de la philosophie*, Hans-Georg Gadamer's "Praise of Theory," and Jacques Derrida's strenuous efforts to salvage the educational benefits of philosophy in all its forms. The main emphasis, however, is on Heidegger's writings in this field—his *Beiträge* (1936), *Besinnung* (1938/39), and *Was heisst Denken?* (1951/52)—which with growing urgency and intensity sought to uphold the importance of humanizing mindfulness in its confrontation with the sway of technological machination and makeability (*Machenschaft*). By way of conclusion, I extend this confrontation to the contemporary global arena where we find the conflicting agendas of unilateral calculating control, on the one hand, and of mindful cooperation, on the other.

Chapter 2 turns to the issue of mindful praxis. Here, contemporary "action theory" has transformed, or rather perverted, human action into the application of precepts of calculating rationality, an application motivated by the pursuit of self-centered goals. In contrast to this conception, the chapter turns to three perspectives which illustrate features of genuinely mindful praxis: those of Heidegger, of Raimon Panikkar, and of classical Indian and Chinese thought. In the case of Heidegger, the focus is placed on his view of action not as a self-propelled "project," but as an opening of the agent to what is encountered or what discloses itself, an opening which does not seek to appropriate the encountered but rather respects its internal dignity and thus "lets it *be*." In some of his writings, Heidegger presented this "letting-be" as the basic or "primordial" praxis, that is, a mode of action which undergirds and sustains all properly mindful human conduct. Turning to Panikkar's religiously inspired philosophy, the chapter recounts his resistance to the notion of God as a calculating omnipotent super-agent reducing human beings to either fatalism or despair. Preferring to see the world as an ongoing disclosure inspired by the "rhythm of Being," Panikkar renews St. Bonaventure's notion of "*creatio continua*" placing humans in the role of co-creators rather than passive victims or self-seekers. Intimations of "letting-be" and co-creation are also present in the Indian legacy of *karma yoga* and the Daoist expression "*wu-wei*." In terms of the *Bhagavad Gita*, *karma yoga* designates a non-self-serving conduct unattached to the fruits of action and fully mindful of the world's well-being. Although frequently misinterpreted as an equivalent of liberal "*laissez-faire*," the Daoist *wu-wei* means a midpoint between action and inaction, where the agent has overcome his/her selfish pursuits and, thus liberated, is able to "wander" freely without constraints.

The next chapter ponders a difficult situation: where thinking encounters its own limit or abyss at the edge of thought. At this point, mindfulness—in the sense of any "fullness of mind"—has to give way to self-abandonment or "mind-fasting" which relinquishes any trace of appropriation or positive knowledge. What this fasting brings to the fore is the inevitable inter-lacing or interpenetration of being and non-being, of positing and nihilation, of life and death. Chapter 3 turns first to Heidegger's attempt to grasp the actively disclosing quality of nihilation beyond the confines of logical negation ("*das Nichts nichtet*"). Pursuing this exploration into post-metaphysical or "meta-metaphysical" terrain, his later writings link nihilation with the Being-question itself, by tracing nothingness (*das Nichts*) to Being's self-withdrawal, sheltering and occlusion. In a

Introduction

parallel manner, by adopting the ontological "difference" (between Being and beings), theologian Paul Tillich discovered in divine Being a dimension of withdrawal and sheltering which can be called the "non-being" of God or (with Jacob Böhme and Meister Eckhart) the "godhead beyond God." Through sheltering of this kind, the divine—far from sinking into sheer negativity—is able to repulse instrumental appropriation and abuse. Along similar lines, classical Buddhist and Daoist thought has built around ultimate Being the impenetrable barrier of "emptiness" (*sunyata, wuji*) foiling human manipulation. Derailing any kind of positivism, the barrier shatters the illusion of the self-contained identity of particular things (*svabhava*), releasing them into the freedom of the "Way" (*dao*).

According to Hans-Georg Gadamer, "Being that can be understood is language."[3] Attentive to his counsel, chapter 4 explores the relation between mindfulness and language (the latter term taken in a broad sense beyond purely linguistic expressions). Mindfulness in this domain means attentiveness not so much to empty chatter, gossip or public media, but rather to the deeper integrity of language, its potentially disclosing quality. Unfortunately, attentiveness of this kind is or has become exceedingly rare in our fast-paced and noisy age; it is as if a bad spell has managed to numb human ears and hearts. The chapter discusses three exemplary contexts in which this spell has been recognized and successfully dispelled. The first context is "Arabian Nights," the collection of tales which princess Sheharazad narrates to king Shahrayar in the course of successive 1001 nights. As becomes clear, the narration of stories is not only a matter of entertainment but a life-saving exercise—because the king, initially bent on killing, in the end is so enchanted by listening to the tales that he grants Sheharazad her life (and being). The second context is recent German theater where the language of classical playwrights (like Schiller and Kleist) is often completely debased to please the depraved tastes of an audience. Here a thoroughgoing "ear-fasting" (akin to mind-fasting) has to take place to attune ears again to the inner splendor and disclosive quality of classical plays. Fortunately, as I acknowledge, some ear-fasting of this kind is happening in a few contemporary stage performances. The third context is that of recent poetry, with a focus on the works of Georg Trakl. What is surfacing in this poetry—as interpreted by Heidegger and other thinkers—is a thoroughgoing mind- and soul-fasting which allows poetic language to regain its inner luminosity, thus revealing its releasing and redemptive potential.

Poetic language is one art form which illustrates the combination of disclosure and sheltering of meaning required for contemporary mindfulness. A similar combination, however, can also be found more generally in contemporary art, especially in visual art (paintings, architecture). As chapter 5 tries to show, contemporary painting has struggled to liberate itself from the stranglehold of modern "aesthetics": more specifically from the antinomies (deriving from Descartes) between subject and object, mind and matter, seeking instead to participate in the ongoing disclosive sheltering of nature (or Being) itself. The point of this struggle has been articulated by a number of philosophers as well as leading painters. The chapter starts with a discussion of Heidegger's famous lectures on "The Origin of the Work of Art" with their emphasis on the counterpoint between disclosure and concealing, between world and earth, and ultimately between truth (in the sense of *aletheia*) and un-truth. After pursuing Heidegger's thinking on the topic in some of his later writings, the focus of the chapter shifts to the works of Paul Cézanne and Paul Klee and to mindful philosophical interpretations of these works. The accent here is placed first on comments by Heidegger (who was closely preoccupied with both painters throughout his life) and then by Merleau-Ponty and by Klee himself. The chapter concludes with reflections on the relation of "world and earth," and on the significance of this counterpoint in the context of the contemporary "worlding of the world."

The theme of chapter 6 is mindfulness in its relation to history and temporal change in general. Here it is important to remember that Being is not a static concept but an unfolding disclosure that happens in time. Such happening is not reducible to clock-time measured by calculating rationality; nor is it akin to an abstract eternity located beyond time and space. Mindful of the meaning of its existence, *Dasein* participates in this happening in its own timely fashion—which can be called an existential or "lived" time where different temporalities (past, present, future) intermingle and where life and death, finitude and infinity intersect. The chapter takes its departure from Heidegger's reflections on time and temporality, reflections found in some early lectures and in *Being and Time*. From his perspective, temporalities are not distinct blocks of time, but rather inter-connected experiences, in the sense that the past is not simply gone but present in recollection, while the future is not simply blank (nor a designable project), but anticipated as impending. What this view of temporality debunks are two major types of derailments: first, the retreat into a mummified and dogmatically reasserted past (as can be found in nostalgic fundamentalisms); and secondly, the radical erasure of the past in favor of willfully constructed ideological "projects" or marching orders. Supported by kindred statements of Walter Benjamin and Theodor Adorno, Heidegger's mindfulness counsels against these derailments—a counsel which is particularly important in our "time" of globalization and cross-cultural learning.

The concluding chapter 7 turns to the contemporary process of globalization and, in this context, to a new mode of mindful inquiry which has come to be known as "comparative or cross-cultural political theory (or philosophy)." The chapter addresses three main issues or queries. First of all (and harking back to earlier discussions): what is mindful thinking or theoretical mindfulness in the particular area of politics and political thought? As previously indicated, mindful thinking involves not the dogmatic assertion but the engaged quest or search for truth and wisdom—which, in the practical-political arena, translates into the search for justice and the "good life." Secondly, what is the meaning of "comparison" or comparative inquiry? Here, the chapter distances itself mainly from apriori or transcendentalizing approaches which, adopting a "view from nowhere," seek to provide an overview of the entire range of different cultural traditions and perspectives. Resisting the lure of an abstract universalism—and also of a "global theory" too easily coopted by hegemonic or imperialist agendas—the chapter favors the outlook of Raimon Panikkar who, in several of his writings, has linked comparative studies closely with dialogical praxis where learning proceeds from mutually engaged encounters. The final question—why comparative cross-cultural studies today?—receives a ready answer by a simple glance at some of the immense dangers and catastrophes of our age: two World Wars, genocide, ethnic cleaning, terror wars. As an antidote, to break the spell of these disasters, I invoke in the end Schiller's "Ode to Joy" as celebrated in Beethoven's Ninth Symphony.

The papers in the Appendix flesh out some aspects mentioned in the preceding chapters. The paper titled "Robots and *Gestell*" takes up the issue of the tension between calculating rationality and mindfulness, but shifts the accent to the steady advances of technology in our world. Taking its bearings from some of Heidegger's writings in this field, the essay shows how—far from being a mere gadget or utensil—technology today functions as a powerful constellation (*Gestell*) which "enframes" human life and, transgressing usability, proceeds to use or abuse human beings, ultimately rendering them useless. The remedy, in Heidegger's account, is not a Luddite revolt but a mind-fasting preparing for human releasement or "*Gelassenheit*." The paper on "Orthopraxis" continues the earlier discussion of mindful praxis, reticence, and sheltered disclosure. Here the focus is on the relation between theoretical cognition and human praxis, an issue stirred up by the Indian social theorist Ananta Giri in some of his recent writings. In opposition to claims of absolute knowledge as well as relativistic denials, Giri draws attention to the *Bhagavad Gita* where *karma yoga* is presented precisely as a thoughtful or mindful praxis geared toward the performance of rightful conduct. He also refers extensively to the teachings of recent "liberation theory" where the emphasis is shifted from orthodoxy to *orthopraxis*, that is, the pursuit of social justice guided by religious faith. The final paper criticizes contemporary agendas of world empire as forms of "pseudo-theocracy" predicated on the fusion of omniscience and omnipotence, and as barriers to the advent of "God's kingdom" in the sense of a just and peaceful life on earth.

NOTES

1. Frequently, "mindfulness" is instrumentalized as a means to lead more effective and productive lives as well as to develop "leadership" qualities. Thus, in America, many businesses, law firms, counseling centers, government agencies, and prisons provide training sessions for employees or participants in mindful meditation. In a similar vein, "coping strategies" is an example of a training program used by the U.S. Armed Forces for its personnel. Compare, for example, Ronald D. Siegel, *The Mindfulness Solution: Everyday Practices for Everyday Problems* (New York: Guilford Press, 2010); M. Carroll, *The Mindful Leader: Ten Principles for Bringing Out the Best in Ourselves and Others* (Boston: Trumpeter, 2007); R. E. Boyatzis and A. McKee, *Resonant Leadership: Renewing Yourself and Connecting with Others Through Mindfulness* (Boston: Harvard Business School Press, 2005).

2. Martin Heidegger, *Gelassenheit* (Pfullingen: Neske, 1959), p. 20; *Discourse on Thinking*, trans. John M. Anderson and E. Hans Freund (New York: Harper & Row, 1966), p. 52.

3. Hans-Georg Gadamer, *Wahrheit und Methode*, 2nd ed. (Tübingen: Mohr, 1965), p. xxi; *Truth and Method*, trans. Joel Weinsheimer and Donald G. Marshall (2nd rev. ed., New York: Crossroad, 1989), p. x.

ONE

Mindful Thinking

The Future of "Theory"

In India, the so-called "Forum on Contemporary Theory" has been celebrating this year (2014) its Silver Jubilee, having come into existence in 1989. As a long-time participant in this Forum, the celebration was for me an occasion to reflect again on the meaning of theory or theorizing, its fortunes and misfortunes, and also on some recent dangers and derailments. As we know, the term "theory" comes from the Greek *"theoria"* which meant the practice or attitude of "looking" or "gazing at," a noun derived from the verb *"theorein"* meaning "to consider, to regard, to ponder, to view or see." Someone engaged in the practice of *"theorein"* was called a *"theoros,"* a looker, a seer, a spectator. The Greek noun *"theoria"* has usually been rendered in Latin as *"contemplatio"* which carries a similar meaning.[1] In its long history, no doubt, theory has gone through many phases and been assigned different connotations or shadings. Perhaps, in a rough overview, one can distinguish between a premodern, a modern, and a postmodern or impending phase. For present purposes, I find this sequence helpful.

HISTORICAL COMMENTS ON THEORY AND PRAXIS

As noted, in the original Greek sense, the theorist or *"theoros"* was an onlooker or spectator. However, was theorizing or *"theorein"* simply a spectator sport in the modern sense of that term? Here we have to take a closer look at looking or seeing. What the theorist in the Greek sense was looking at was not some random phenomenon but rather the "cosmos" or "Being" in its manifold appearances. This cosmos, in turn, was not just

9

a cause-effect nexus but a meaningful whole, a whole permeated by, and oriented toward a *telos*: a *telos* expressed by Plato as the loadstars of truth, goodness, and beauty. And since the whole was really a whole (excluding nothing), it necessarily also included the viewer or seer as a being oriented toward the same *telos*. Hence, the viewer was not merely an onlooker or bystander (on the outside), but a participant in the great disclosure or epiphany of the cosmos.

As one can see, this kind of "*theorein*" was still far removed from the modern dichotomy of theory and praxis, of knowing and doing. Rather, theorizing was itself a praxis: the praxis of participating in the movement toward truth, goodness, and beauty. In order to participate properly in this movement, the theorist had to undergo a certain seasoning, purification, or transformation, a seasoning enabling the theorist to be whole-heartedly devoted to, or focused on, truth, goodness, and beauty. Just as in order to physically see one needs to have eyes, in order to hear sounds one needs to have ears, so in order to observe, regard, ponder what is genuinely "real" one has to have a heart-mind which is fully attentive or attuned to "Being" expressed traditionally as the *telos* of thinking.

Here, theorizing and practicing are closely linked. To put things more sharply: theory in the Greek sense requires a combination of interest and disinterest. The theorist needs to be fully engaged and completely, even urgently *interested* in the quest for truth, goodness, and beauty; but at the same time, the theorist needs to bracket selfishness and to be *disinterested* in the pursuit of his/her own particular "good" or advantage. Much later in history, the French thinker Maurice Merleau-Ponty will say that theory or philosophy is a "limping" enterprise, and this is correct because it limps between interest and disinterest, between whole-hearted involvement of self and whole-hearted self-abandonment.

Later on in Western history, the Platonic *telos* of theory—which, by the way, Aristotle never discarded—was merged with, or supplemented by, Christian faith and its intrinsic yearning to perceive or see the divine "face." Despite many struggles between philosophy and theology during the Middle Ages, the shared assumption of both was that proper seeing, understanding or "knowing" depended on the genuine preparation or attunement of the seer or seeker, that is, on a certain practical orientation and habituation. While for medieval philosophers the practice involved the inner attunement to the Platonic triad—sometimes reinterpreted along more spiritual, neo-Platonic lines—for theologians the practice included prayer, meditation, and, of course, divine assistance. It was in the latter context that the fascinating idea of "*theosis*" was formulated, that is, the idea of the ongoing "divinization" of the faithful as a requisite preparation for a deeper understanding of the essence of faith. In Western Christianity, this idea—most prominent in Eastern orthodoxy—finds a parallel in mystical contemplation and in Saint Bonaventure's notion of the "*itinerarium mentis in Deum.*"[2]

This older conception of theory and theorizing came to an abrupt end with the onset of Western modernity. The major change that happened was the replacement of a teleological cosmos with an indifferent universe denuded of meaning and purpose; coupled with this change was the transformation of the theorist from a participant into a detached onlooker or analyst. This shift was most prominently and most dramatically announced in René Descartes's philosophy, especially in his bifurcation between the *cogito* and the rest of the world. With this bifurcation, the thinking ego was in fact expelled from the world and thus turned into a pure spectator or onlooker; the goal of the ego was to know itself and the world, but knowledge was no longer predicated on existential participation. The Cartesian move was buttressed by Francis Bacon's inauguration of the "new science" inspired by the motto "knowledge is power"—a motto where human power over nature replaced the earlier accent on attunement, transformation and any kind of "*theosis*." Carried forward by Enlightenment philosophy, the Cartesian and Baconian formulas became the distinctive hallmark of Western modernity—although one should not forget the many currents and counter-currents inhabiting the modern era. Even a great philosopher like Immanuel Kant had to invest great effort to reconcile or re-connect pure (detached) theory and ethical praxis—and his effort was not entirely successful. As he wrote soberly at one point: "A theory which concerns objects of perception [seeing, viewing] is quite different from one in which such objects are represented only through concepts, as with objects of mathematics and [abstract] philosophy. The latter objects can perhaps quite legitimately be *thought* of by reason [the *cogito*], yet it may be impossible for them to be given [in experience]."[3]

Since the heyday of the Enlightenment, Western modernity has moved steadily and relentlessly in the direction of pure knowledge and pure science, that is, a spectatorial vista far removed from engaged practice. Every so often, to be sure, this trajectory was disrupted by the upsurge of various kinds of fundamentalism (national, ethnic, religious)—but without altering the course of "reason." Even counter-currents within the ambit of science—like Einstein's relativity theory—were not able to affect this course in any lasting way. The goal point of the modern trajectory was to reach a detached vantage point from which the whole universe could be cognitively grasped and subjected to technical control. In Baconian and Cartesian terms, the goal was to establish a central all-knowing *cogito* whose absolute knowledge of everything would give it absolute power.

In near-prophetic manner, this idea of central spectatorial power was anticipated in Jeremy Bentham's notion of the "*panopticon*"—whose stark political implications were much later discussed by the French philosopher Michel Foucault.[4] In our own time, the program of a "*panopticon*" has been implemented in the form of massive surveillance centers established by global hegemonic power, centers which allow the gathering of data about everything happening anywhere in the world. As always, science and technology are closely linked: global knowledge entails global control—evident in the ability to eliminate all obstacles to global "security" (preferably via remote control).

"ÉLOGES DE LA PHILOSOPHIE"

Fortunately, the contemporary situation is not as absolutely bleak as these comments indicate. During the past century, counter-currents which had been marginalized by the dominant culture have been able to gather strength and to usher in new developments. A first initiative able to break through the hard shell of the theory-practice split was John Dewey's philosophical pragmatism, a perspective which insisted that truth and goodness could only be found through participatory engagement. The major breakthrough, however, happened with the movement of "phenomenology," a movement which upheld the need of a new manner of "seeing' or "looking" at appearances, that is, a new practice of *"theorein."* Basically, in opposition to the self-sufficiency or self-enclosure of the *cogito*, phenomenology insisted on the opening or opening-up of the "seer" or theorist to the meaning or significance of phenomena. As formulated by its founder, Edmund Husserl, the movement (it is true) still carried some traces of Descartes' transcendental spectatorship; however, these traces were corrected by Husserl's later turn to the "life-world" and the theorist's own involvement in that world. Most importantly, the initiative reached its full potential in the next generation: especially among French phenomenologists and among German practitioners of phenomenology and hermeneutics.

Among the former group, the most innovative and promising mode of "looking" and "theorizing" was introduced by Merleau-Ponty. As previously indicated, it was Merleau-Ponty who characterized philosophy as a "limping" enterprise because it involves both the complete devotion of the thinker to truth and goodness and a complete unselfishness (which also means that truth cannot be possessed). The description is found in his famous *Éloge de la philosophy (In Praise of Philosophy)* delivered as inaugural lecture at the Collège de France in 1953. In Merleau-Ponty's presentation, "limping" refers not only to the conjunction between interest and disinterest, but also to the difference or differential embroilment of presence and absence, of location and dislocation. In devoting himself/herself to the truth, the philosopher is not placed outside humanity into a never-never land of perennial verities; nor can he/she find an abode within the "cave" of a comfortable social community. Rather, philosophy is always underway, on a quest. In Merleau-Ponty's words: "A philosophy of this kind understands its own strangeness, for it is never entirely in the world, and yet never outside the world. . . . Philosophy cannot be a tête-àtête of the philosopher with the 'true.' It cannot be a judgment given from on high on life, the world, history, as if the philosopher *was not part of it*— nor can it subordinate the internally recognized truth to any exterior instance of it. It must go beyond this alternative."[5]

For Merleau-Ponty, the thinker who best exemplified this tension (at least in the Western tradition) was Socrates. As he points out, Socrates did not exit from, or contemplate the destruction of, his city (he never left Athens); nor did he find a resting place in it, accommodating himself to reigning beliefs. "The life and death of Socrates," he writes, "are the history of the difficult relations the philosopher faces . . . with the gods of the city, that is to say, with other human beings and with the fixed absolute ideas whose image they extend to him." What renders the Socratic position so baffling is that he is neither inside nor outside, neither fully located nor fully dislocated: "He teaches that religion is true, and he offered sacrifices to the gods. He teaches that one ought to obey the city, and he obeys it from the very beginning to the end. He is reproached not so much for what he does as for his way of doing it." Basically, Socrates demonstrates the peculiar quality of philosophizing or "theorizing": its being in the world, its participating in the city, while maintaining its "apartness" (*Abgeschiedenheit*). Socrates, Merleau-Ponty concludes, "works out for himself another idea of philosophy: It does not exist as a sort of idol of which he would be the guardian and which he must defend. It rather exists in its living relevance to the Athenians, in its present absence, in its obedience without respect."[6]

Some twenty-five years after Merleau-Ponty's inaugural address, the German philosopher Hans-Georg Gadamer presented in Bonn a public lecture with the title *"Lob der Theorie,"* which means: *Éloge de la théorie* or (in the English translation) *Praise of Theory*. Without citing the work of his French colleague, Gadamer's lecture in many ways parallels and corroborates Merleau-Ponty's text. Without invoking phenomenological language, the lecture recalls the early meaning of theory or *"theorein"*: namely, "seeing what is." The target of this seeing is not simply an empirical "fact," but rather the meaning or significance of "what is," something which cannot be reached without existential participation. Thus, Gadamer says, the target is "hermeneutical" (interpretive), which means "that it is always referred back to a context of supposition and expectation, to a complicated context of inquiring understanding" (that is, a lived context). Participation here is not merely an "individual momentary act," but a way of life, a manner of "being present" in the full sense. As in Merleau-Ponty's case, presence here involves a peculiar juncture of presence and absence: presence in the sense of attunement to what is, and absence in the sense of a lack of selfish desire. If it is properly human, Gadamer asks, is theory not "a looking away from oneself and looking out toward the other, disregarding oneself and listening for the other?" If this is so, then we discover here a unity of knowing and doing that is a general human possibility and duty: "Disregarding oneself, regarding what is: that is the emblem of a cultivated, I might almost say a divine, disposition."[7]

As a keen student of the history of philosophy, Gadamer acknowledges that this sense of theory has atrophied or disappeared in Western modernity, being replaced by abstract scientific cognition. The primacy of the *cogito* in modern thought, he observes, entails ultimately "the primacy of method"—which means in turn that "only what can be investigated by method is the object of science." When modern science took the path of method, he adds, "it ignited a true explosion that burst apart not only the Middle Ages' geocentric view of the world, but also its theocentric one." The pursuit of knowledge *more geometrico* was elevated to the standard way of knowing reality. The connection between knowing and doing or between thought and practice was shunted aside—unless the latter was reduced to a mere instrumental or technical application of science. However, Gadamer asks here: "Is there perhaps more to theory [*theorein*] than what the modern institution of science represents to us? And is practice, too, perhaps more than the mere application of scientific formulas?" To overcome this truncated conception of theory and practice, he turns like Merleau-Ponty to classical thought. As a trained student of the classics, he takes as his guide not only Socrates but also Plato and Aristotle. "In the paradox of the philosopher king," and it must remain a paradox, he notes, "Plato articulated a lasting truth: namely, that being fit to rule over others or to carry out any official function can mean only knowing what is [ethically] better. . . . So the ideal of the 'theoretical' life does have practical-political significance." Hence, for Plato, "theory and politics remained indissolubly [though paradoxically] united." And Aristotle carried on this union, seeking "to legitimize the proper balance of both, the ideal of the practical-political life and the integrity of the theoretical one."[8]

Roughly at the time of Gadamer's lecture, French philosopher Jacques Derrida issued an eloquent plea in favor of broad-based philosophical education (one of his persistent public engagements). The plea was formulated in the course of an interview which was subsequently published in 1981 under the title "In Praise of Philosophy." Derrida at the time was a member and prominent supporter of a group of intellectuals called GREPH (*Groupe de Recherches sur l'Enseignement Philosophique*), a group which in prior meetings had advocated the expansion of philosophical instruction and also the eventual construction by the state of an international college of philosophy. The interview—and the entire series of discussions or debates sponsored at the time by GREPH—was characterized by a sense of urgency or crisis: the sense of a "crisis of philosophy," the latter seen not as scientific method or epistemology, nor as the rehearsal of perennial verities, but as the "limping" enterprise mentioned by Merleau-Ponty, situated at the cusp of location and dislocation. As Derrida pointed out, the enterprise which was in need of defense was not so much professional philosophy as a recognized discipline with its routinized canon, but rather "something like thinking at the limits of philosophy . . . in novel forms, on contents that are new and still little or poorly represented in the current distribution of the fields of teaching." It was this latter kind of thinking or theorizing (*theorein*) which was under assault today, not only by governmental edicts and restrictions, but by a whole battery of forces, including "the powerful constraint of the market, techno-economic imperatives, and a certain concept . . . of immediate adaptation to the urgencies of productivity in national and international competition." Using broad strokes one could summarize these forces under the labels of "technologism," "productivism," and reductive "positivism."[9]

From Derrida's angle, modern Western philosophy (especially in its "analytical" guise) had become subservient to, or an "under-laborer" of, science—while the latter had become subservient to the advances of technology and economic productivity. From the vantage of this subservience, innovative philosophical thinking was simply "not profitable, not sufficiently 'productive' (*performant*)." Some extreme defenders of scientism or technologism had even gone to the length of propounding the need "to evacuate or eliminate philosophy and everything that did not respond to the criteria of productive 'performance', the so-called 'social needs'." Faced with this technical onslaught, Derrida did not advocate the abolition of science or technology, but a broader, richer conception of philosophizing or thinking (*theorein*). As he pointed out, the term "philosophy" today names at least two things. On the one hand, the term refers to a rich tradition of texts and arguments, spread out into such fields as metaphysics, epistemology, practical philosophy and the like. But on the other hand, the term is also "rightly associated with every 'thinking' that no longer lets itself be determined by techno-scientific or cultural programs, that troubles them sometimes, interrogates and affirms them, yes, affirms beyond them, without necessarily opposing or limiting them in a 'critical' mode" (what is called "deconstruction" being one such critical affirmative mode). Taken in this second sense, philosophizing or thinking (*theorein*) would be "that mobile non-place from which one continues or begins again, always differently, to ask oneself what is at stake in technology, the positivity of the sciences, production, yes, and above all, productivity."[10]

As one should add, the "non-place" or dislocation of philosophy meant for Derrida also that it could not be simply identified with "Western" philosophy, but had to be placed in the interstices of West and non-West, of localism and globalism. This point was forcefully made in a text published by UNESCO in 1997 and titled (in English translation) "The Right to Philosophy from the Cosmopolitan Point of View." The text took its point of departure from Kant's famous essay "Idea for a Universal History with a Cosmopolitan Purpose." In his comments, Derrida ac-

knowledged Kant's effort to exit from a narrowly nationalistic or ethno-
centric framework, but at the same time critiqued this effort for being too
blandly "universalistic" in the Enlightenment sense—and precisely on
this account for slipping back into a Eurocentric teleology. As he writes,
Kant's text might be read as "the expression of a confident optimism and,
above all, an abstract universalism"—an outlook which is beset with profound difficulties, mainly because of its cultural bias. Although overtly
cosmopolitical in spirit, the essay—in Derrida's view—is "the most
strongly Eurocentric text possible, not only in its philosophical axiomatic
but also in its retrospective reference to Greco-Roman history and its
prospective reference to a future hegemony of Europe." As an antidote to
this trajectory, Derrida's text stresses the need to "take into account and
delimit the assignation of philosophy to its Greco-European origin"—
though not in the mode of a simple reversal which would localize philosophy *elsewhere*. The task, he insists, is rather "to displace the fundamental
schema of this problematic by going *beyond* the old, tiresome opposition
between Eurocentrism and anti-Eurocentrism." This means acknowledg-
ing that philosophy is no more "assigned to its [cultural] origin" than it is
simply or abstractly universal or cosmopolitical. [11]

   According to Derrida, what is emerging today are philosophical ave-
nues that cannot be locked into the old binaries of "appropriation and
alienation," of cooptation and radical rejection. Not only are there novel
or other ways of philosophizing, but "philosophy *is* the other way," for it
is never finished but always *en route*. Summarizing his main argument,
Derrida lists three points designed to combat any hegemonic confine-
ment of philosophy. The first point involves a certain cosmopolitan agenda not tied to Eurocentric blinders. Philosophy or the "right of all peoples
to philosophize," he says, requires "the reflection, the displacement and
deconstruction of all hegemonies" (including those of Continental and
analytical philosophy). This requirement has important implications for
language (second point). For, if the hegemonic and all-powerful role of a
given language—"I mean English"—can serve as vehicle for philosophical communication, then philosophy demands by the same token "that
we liberate ourselves from the phenomena of dogmatism and authority
that this language can produce." The point here is not to remove philosophy from language and even from vernacular idioms, but rather to encourage—in a "nonfinite multiplicity of idioms"—the flourishing of discourses that are "neither particularistic and untranslatable nor transpar-
ently abstract and univocal." This point is particularly relevant for hegemonic discourses, like those of contemporary sciences, to the extent that
they bow to the univocal dictates of technocracy and performativity. The
"hypothesis" that Derrida's text in the end submits for discussion is that
the general right to philosophize involves "not only a politics of science
and technology but a politics of *thinking* (*pensée*) that yields neither to
positivism nor to scientism nor to epistemology" and thus recovers its

"irreducible autonomy." The text concludes with a plea (third point), in view of the present dangers, for a broad-based and indeed cosmopolitan education in philosophy.[12]

THINKING AS A HUMANIZING PRAXIS

As is well known, the three writers discussed so far have all been influenced, to a greater of lesser degree, by the work of the most prominent recent philosopher of philosophy or thinker about thinking: Martin Heidegger. Throughout his life, Heidegger's work has always revolved around the challenge and importance of philosophizing or thinking. The central point—one recalls—of his chef d'oeuvre *Being and Time* (1927) was the renewal of "the question of Being," a renewal which requires careful attention and thinking. The thinking or questioning involved here operates in two directions. On the one hand, human beings are compelled to raise the "question of Being," by asking "what does it mean for us to be"; but at the same time, Being calls humans into question, by depriving them of any fixed or preordained script or any pat recipe for being or living.

To be sure, in *Being and Time*, the questioning is embedded in a complex network of ancillary topics and concerns, such as the issues of understanding, language, care for others, and temporality; yet, none of these topics would get off the ground without the careful thinking required by the question of Being. Despite many subsequent modifications—and even the fabled radical *"Kehre"*—Heidegger never abandoned his basic starting point. Thus, almost forty years later, in one of the so-called "Zollikon Seminars," he stated that "in *Being and Time* the question of Being (*Seinsfrage*) determines everything, that is, the question how Being discloses itself in time." And another decade later, in May 1976 (thus shortly before his death), he wrote in a message to a conference in Chicago: "The question with which I send my greetings to you is that single question which I have persistently tried to ask in a steadily more questioning manner: It is known as 'the question of Being'."[13]

As one needs to realize, Heidegger initially raised this question not as a solitary recluse, but as an academic teacher of philosophy—in fact, as one widely recognized as one of the best teachers in this field. Long before Derrida, Heidegger recognized the crucial importance of philosophical education and teaching—a teaching concerned not with an esoteric or exotic subject matter but a practice touching the very core of human being. Repeatedly in his career he offered lecture courses in this area; one such course was presented in Freiburg in 1928/29 under the title "Introduction to Philosophy" (*Einleitung in die Philosophie*). As Heidegger made it clear right at the outset, "Introduction" here cannot mean the mere gathering or piling up of data *about* philosophy, such as the historical sequence of philosophical "systems" or doctrines and the differences among systems. Nor can "Introduction" mean a guided tour among the so-called branches of philosophy, like logic, ethics, and aesthetics, and their historical development.

What such data gathering accomplishes is an expansion of information or cognitive knowledge, but without existential engagement or an inducement to philosophize. In Heidegger's words: If, at the end of the semester, students have completed such an "historical and systematic overview," they are "the proud owners of sundry modes of information," but they have learned nothing about "philosophizing" seen as a praxis or practical endeavor. As he adds, the entire notion of an "Introduction" is misguided if it implies a movement from a place outside of philosophy to an insider's perspective. It is misguided because philosophizing is not a rummaging in ideas but an existential and transformative undertaking. In the poignant words of the lecture course: "Even if we do not explicitly know anything about philosophy, we are already *inside* philosophy because the latter is something endemic in that we are always already philosophizing." Then follows the still more lapidary formulation: "To be human (*Menschsein*) means to think or philosophize. Human *Dasein* as such—in its very nature and not occasionally or accidentally—is anchored in philosophy." [14]

In most human beings, to be sure, philosophy exists initially only as a latent potential or in a dormant state requiring awakening; to this extent, "Introduction" means an attempt to actualize a dormant possibility. In Heidegger's words: "Introducing now means to mobilize (*in Gang bringen*) philosophizing, to allow philosophy to become in us an actual happening (*Geschehen*)." When this occurs, human beings are drawn into the pull of a quest: the relentless quest for truth, goodness, and beauty. This quest can also be construed as a mode of emancipation or liberation: liberation *from* extraneous preoccupations or addictions, and liberation *of* the deepest human potential and task: that of the "humaneness" (*Menschsein*) of human beings. As Heidegger makes it abundantly clear, mobilizing the freedom to philosophize does not at all mean to reject or jettison the historical legacy of philosophy. Even less does this freedom sanction a retreat into self-centered subjectivism or solipsism. "The liberation of philosophy in human *Dasein*," the lecture course insists, "has nothing whatever in common with a psychological or egotistical narcissism (*Selbstbegaffung*)." Contrary to the modern view of the centrality of the *cogito*, philosophizing as a radical quest reveals that human being is "in its core ek-centric," that by its very nature "it can never occupy the center of beings" or "possess itself." By entering the pull of thinking or philosophizing, human *Dasein* is in fact "catapulted out of and beyond itself" and thus is open to the appeal of Being and the truth of Being. [15]

Seen as an awakening and existential quest, philosophizing clearly has a transformative and, in a sense, pedagogical quality. Heidegger's lecture course draws the attention of readers and students explicitly to the connection between thinking or philosophizing and the Greek notion of *"paideia"* meaning educational guidance, formation or *"Bildung."* If, he states, human *Dasein* is animated by the "question of Being" and oriented toward the "understanding of Being" (*Seinsverständnis*), then philosophizing inevitably has the character of *"paideia."* Understanding Being, from this angle, is not sudden or instantaneous but requires sustained labor, a labor that, in turn, has to be propelled by an existential desire or inclination, in Greek *"philein"*—which discloses the inner affinity between *"paideia"* and the original Greek meaning of "philo-sophia" (love of wisdom). In the words of the lecture course: "Understanding requires a sustained effort which, from the outset, has to be nurtured by an original inclination or sympathy (*Neigung*) toward things. This sympathy, this inner friendship with beings is meant by the Greek *'philia'*—a friendship which, like very genuine liking, has to struggle with and for the target of its love." With this accent on loving and struggling, the notion of philosophizing is clearly far removed from detached spectatorial observation or a mere conceptual analysis, and inserted into a mode of praxis and existential humanization: "Understanding is not something that can happen without engagement (*Zutun*); rather, it needs to be lifted up into the freedom of human existence in order to be actualized."[16]

During the following decade, dark clouds began to gather over Europe—clouds from which Heidegger could not and did not shield himself completely. In response to these developments his writings increasingly exhibited the desire to profile more sharply the meaning of philosophizing, and especially to distance his own notion of thinking from a purely cognitive-spectatorial stance which, under the aegis of the *cogito*, tends to subject the entire world to human control. Initial steps in this direction were taken in two major works written during the 1930s (but published much later): *Contributions to Philosophy* (*Beiträge zur Philosophie*, 1936), and *Mindfulness* (*Besinnung*, 1938/1939). The first volume offered a dramatic sketch of the successive steps that need to be taken in order to overcome the dominant mode of (Western) philosophizing, a mode aiming at the cognitive grasp and total knowledge of the world (omniscience) coupled with the total technical control of everything (omnipotence). As Heidegger observed, modern philosophy has endeavored to be a "scientific philosophy," a theorizing grounded in anthropocentric cognition oriented toward "unitary and systematic-mathematical knowledge (*Wissen*)" of the world in the form of apodictic "certainty" (*Gewissheit*). In order to align this knowledge with some cultural or value-laden commitments, modernity has fostered the growth of totalizing "world-views" or "world pictures" (*Weltanschauungen*) designed to furnish their adherents with action platforms or marching orders. What comes to the fore in these world-views and programs is not a genuine form of praxis, but rather an anthropocentric will to knowledge or will to power expressing itself in instrumental fabrication and the glorification of "machination" or the "makeability" (*Machenschaft*) of everything. Both scientific philosophy (styled as "epistemology") and world-views are far removed from the domain of genuine thinking where philosophy is seen as "the grounding of the truth of Being in truth itself." Against this background, what is needful today is "mindfulness (*Besinnung*)" regarding the "genuine destiny of philosophy," together with awareness of what "disturbs and disfigures" thinking today.[17]

These comments were developed and fleshed out further in *Mindfulness* a few years later. The text is a fervent denunciation of the modern infatuation with fabrication, technical making, and production (*Machenschaft*)—all rooted in the rootless desire for power or the self-empowering of power. "Machination," Heidegger writes, "means the all-making and all-producing makeability of everything, such that all beings are determined or defined by their oblivion of Being. Differently put: machination is the sole focus on instrumental production in such a way that everything becomes countable or calculable." By rendering everything calculable, machination also makes everything wholly knowable, a knowledge which—in accordance with Bacon's motto—yields total power and domination. Such domination, we read, unfolds in "the explosive and constantly changeable capacity for arbitrary and ever-expanding subjugation and control." The capacity for this control is manifest in modern technology which unleashes human beings into the total "ordering of mass society" while erecting its "technical power over all beings." In recent times this dominion of technology—in its "self-overpowering of power" (*Sichübermächtigen der Macht*)—has reached not only "imperial" but global or planetary dimensions: "One speaks of 'planetary reach' and wants to say that today seizures of power are not only 'total' in a local sense but set their sights at the boundaries of the inhabited globe . . . which means that the planet itself becomes a resource of power and that hence finding a global enemy becomes imperative." Needless to say, this unleashing of power carries with it enormous danger, in fact, the risk of global destruction: machination expands the arena of "spiraling annihilation" (*Vernichtung*).[18]

In opposition to planetary cognitive domination, Heidegger's text introduces an "other kind of thinking" (*anderes Denken*) which, in an approximating way, may be called "mindfulness" (*Besinnung*). As Heidegger writes in an opening section: The issue today is "whether the machination of everything shall overpower human beings and make them into limitless power-seekers, or whether Being would again disclose the grounding of its truth as the need out of which the encounter of God and human beings . . . could again emerge." The pondering of the latter possibility is the task of "mindfulness" seen as the thinking or thoughtful recollection of Being. Such recollection can only arise from the needful experience of the non-apodictic "un-ground" of the truth of Being, a truth by which all beings in the world are infused and illuminated, opening them up to their quest for truth. Without this opening or "clearing" (*Lichtung*) all philosophy remains stranded in the endless imitation and repetition of metaphysical formulas—formulas under whose sway the question of the truth of Being remains "unintelligible." Unfortunately, much of philosophy today is in fact stranded in this manner: by either absconding into abstract doctrines or else by surrendering itself to emotional appeals or ideological marching orders. Only through the thoughtful recollection (*Erdenken*) of the truth of Being can thinking return to its innermost necessity or needfulness. This means that, only when impelled by its own inner need, can philosophy regain a self-understanding or mindfulness of itself strong enough to recapture its own task more originally. Such mindful recollection does not "calculate" benefits and losses but places itself into that clearing where it only "minds" the truth of Being.[19]

After World War II, the issue of thinking and mindfulness became for Heidegger an all-consuming concern; equally strong was the need to profile thinking more sharply against a "calculating" rationality which, in its alliance with fabrication and "machination," tends to imprison human life in a technological cage (*Gestell*). His final lecture courses at the University of Freiburg (in 1951 and 1952) were devoted precisely to this issue. The text of these courses was published later under the title *Was heisst Denken?* a title which has been mis-translated as *What is called Thinking?* The point of the text, however, is not to offer a definition of thinking (or what is commonly so called), but rather to offer an invitation into thinking by exploring what solicits or "calls forth" thinking in the first place.

The starting point of the text is the wide-spread inability to "think," which entails our need to learn how to think—which, in turn, is only possible if we focus our attention on what needs to be thought. In Heidegger's words: "We call the essential character of a friend the 'friendly' (*das Freundliche*). In the same manner, we can call the essence of what needs to be thought about the 'thought-provoking' (*das Bedenkliche*)." But what is that which most urgently needs to be thought about today? In the lapidary formulation of the text: "Most thought-provoking in our thought-provoking time is that we are still not thinking." This verdict, Heidegger admits, may seem far-fetched and even implausible, given the widespread interest in philosophical "ideas" and the sometimes glamorous self-display of philosophers as "thinkers." However, preoccupation with seemingly philosophical matters is by itself no evidence of a readiness to think; on the contrary, such preoccupation may generate a "stubborn illusion" of thinking while totally ignoring what needs to be thought.[20]

Having started provocatively—by speaking about the "most thought-provoking"—Heidegger's text in the following becomes steadily more provocative and disturbing. As he says, the fact that we are still not thinking is due not to an unwillingness to learn or a lack of pedagogical institutes, but to a basic dilemma: the fact that what needs to be thought withdraws itself (and has always withdrawn) from our grasp or appropriation, thereby drawing or pulling us into an abyss. When pondering this abyss or vortex, there is nothing we can hold on to, no reliable banisters, nothing we can empirically observe and analyze—that is, we move outside the boundaries of "science." As Heidegger observes, modern science has erected its imposing edifice on the basis of empirical data and epistemic knowledge—which founders at the abyss of data. In his crisp language: "Science does not think." In fact, science cannot think (in the sense sketched above), and this is its good fortune because this alone enables it to proceed securely on its chosen course. In moving outside this domain—in the direction of the most thought-provoking—there is nothing we can securely describe or analyze, because we are pulled toward something which withdraws and shelters itself. All we can do is gesture or "point" to what withdraws. From this angle, human beings are in their core "pointers" (*Zeigende*) toward what eludes complete epistemic knowledge: "By drawing toward what withdraws and thus pointing into the withdrawal, human beings become human in the first place." But something which in itself is pointing is called a "sign" (*Zeichen*). Heidegger at this point invokes a line from one of Hölderlin's hymns: "We are a sign that is not read (*deutungslos*)."[21]

If pointing means being drawn into an abyss, then the move from epistemic knowledge and calculation to "thinking" is not a smooth transition or progression but rather requires a kind of "leap" (*Sprung*) involving risk. To venture the move, and thereby to learn how to think, students need to find teachers—not professors of philosophy but people who exemplify thinking in their way of life. For Heidegger, as for Merleau-Ponty, Socrates was such an exemplary teacher. "Throughout his life and right onto his death," we read, "Socrates did nothing else but to place himself into the pull of this draft or current, and to maintain himself in it. This is why he is the purest thinker of the West." It is also why he did not write anything, for anyone who turns from thinking to writing is "like those people who seek refuge from a powerful storm in a wind shelter." From this angle, many later Western philosophers appear like refugees. (Heidegger's lecture course discusses in this context Friedrich Nietzsche as an exemplary thinker who refused to seek shelter—a discussion I bypass here). If the move from calculation to thinking involves risk and danger, it nevertheless is not a reckless adventure. Although the storm unleashed by withdrawal is turbulent, its center is calm and peaceful— because ultimately the storm guides human beings into their true humanity. Out of the turbulence, humans receive the capacity of thinking not as an achievement but as a gift—to which the appropriate response is gratitude and thankfulness. Seen from this angle, thinking is closely related to thanking, and also to memory (*Gedächtnis*) and recollection (*Andenken*). In the words of the text: "Pure thanking means that we truly think—think what is truly and really given to thought as thought-provoking."[22]

Roughly during the same period, Heidegger presented a number of public lectures (to which I can only gesture) dealing with similar or related topics. Most relevant in the present context are the lectures "What Calls for Thinking?" (1952) and "Science and Mindfulness" (1953). Building on his lecture course, the first talk differentiates still more clearly between modes of theorizing, especially between science and thinking. Modern philosophy, Heidegger observes, has been largely "scientized" by privileging epistemic cognition. Based on the Cartesian *cogito*, epistemology seeks to gain access to phenomena by placing them as "objects" before the human mind or reason, thereby centerstaging the act of "representation" or rendering present (*Vorstellung*). By contrast, in thinking human beings are available or responsive to what calls upon them or solicits their attention.[23] The topic is deepened in the second lecture through a juxtaposition of science and mindfulness. For Heidegger, science is by no means un-theoretical or anti-theoretical, but its theorizing has the form of epistemic knowledge achieved through calculation and measuring of data. Citing Max Planck's statement that "real is what can be measured," Heidegger observes that the decision over what is "real" and "secure knowledge" lies in the measurability of data (where measuring can take many forms)—a method which has yielded enormous amounts of information and staggering advances in technology. Compared with these advances, mindfulness appears devoid of tangible results and marked by destitution or abysmal poverty. However, as the text says, "the poverty of *Besinnung* is the promise of a treasure whose contents cannot be marketed or consumed or calculated." It is the promise of a gift needed by human beings "for the sake of their humanity."[24]

ANOTHER BEGINNING?

Half a century has passed since Heidegger's late writings discussed above. In the meantime, the global situation has grown still darker and more threatening. In his *What Calls for Thinking?* Heidegger invoked Nietzsche's saying "The desert grows; woe to those harboring desert."[25] In the intervening years, the desert or wasteland has not stopped growing. Coupled with other innovations, the "information revolution" has conjured up the prospect of a total calculation of everything, of a universal *"panopticon"* fulfilling the dream of human "omniscience." Supplemented by the advances in military technology, this prospect yields the vision of human "omnipotence" capable of not only knowing but destroying everything. Although allowing for the possibility of an other thinking, Heidegger's late writings do not mince words about looming dangers. Referring to the statement of a Nobel laureate that human life will soon be "placed in the hands of the chemist," he states:

> We marvel at the daring of scientific research, without thinking about it. We do not stop to consider that a technological attack is being prepared upon the life and nature of human beings compared with which the explosion of the hydrogen bomb means little. For even if the hydrogen bombs do *not* explode and human life on earth is preserved, an uncanny change in the world moves upon us. Yet it is not that the world becomes entirely technical what is really uncanny. Far more uncanny is our being unprepared for this change, our inability to confront mindfully (*besinnlich denkend*) what is dawning in this age.[26]

Heidegger's warnings are sometimes dismissed as the bleak ramblings of a Cassandra; but his comments actually pale when compared with ongoing developments and daily experiences. More serious are accusations that his notion of "thinking" is rooted in the Western philosophical tradition; that it is a solitary individual enterprise; and that it needs to be replaced or at least supplemented by practical reasoning and judgment. The first charge is undeniably correct—but not necessarily in a debilitating sense. Rather than starting from a rationalistic "universalism," Heidegger begins from a distinct location, but then opens it up to dislocation and interrogation; precisely his stress on wayfaring and "open seas" makes possible a cross-cultural and cross-temporal engagement. The charge of solitary individualism flies in the face of his critique of Cartesianism and the accent on human "being-in-the-world" as the matrix for mindfulness and genuine thinking. As for the issue of practical reasoning and judging—surely desirable features or qualities of public life—it seems to me imprudent to separate them from "thinking" in the Heideggerian sense. Given the "information revolution" and the widespread manipulation of opinion by media and ruling ideologies, human beings more than ever need to be liberated or released from the "*idola fori*" of the age. Even the meaning of social and public life cannot be taken for granted, but needs to be reassessed—which can only happen through resort to mindful thinking.[27]

The latter turn, it seems to me, is particularly needful in the contemporary global arena. In this domain, the two basic orientations of thought distinguished by Heidegger surface in the form of two conflicting global agendas: one aiming at the universal and monological domination of the world, the other at the mindful cultivation of interdependence and cooperation. The former seeks to concentrate knowledge and power in a central calculating machine, the latter tries to safeguard, as inobtrusively as possible, concrete phenomena and everyday experiences without imposing on them a universal doctrine. In his "Letter on Humanism" (of 1946), Heidegger delineated in a striking manner the basic character of the thinking needful in our time, highlighting three main features: "rigor of thinking, carefulness of expression, and frugality of words." As he added: "What is needed in the present world crisis is less 'philosophy' but more attentiveness of thinking, less literature but more cultivation of the letter." The conclusion of the text sums up the gist of Heidegger's orientation: "Thinking is on the descent into the poverty of its provisional nature. Thinking gathers language into a simple saying. In this way, language is the language of Being, as clouds are the clouds of the sky."[28]

NOTES

1. For the etymological background see Martin Heidegger, *Einführung in die Philosophie*, ed. Otto Saame and Ina Saame-Speidel (*Gesamtausgabe*, vol. 27; Frankfurt-Main: Klostermann, 1996), 167–170. Heidegger links "*theoria*" also with "*theos*" and "*theia*," and "*contemplatio*" with "*templum*," adding (169) that, in Greek etymology, "*theos*" and "*theoria*" refer basically to the "viewing of divine things."

2. See in this context my "A Pedagogy of the Heart: Saint Bonaventure's Spiritual Itinerary," in *In Search of the Good Life: A Pedagogy for Troubled Times* (Lexington, KY: University of Kentucky Press, 2007), 23–39.

3. Immanuel Kant, "On the Common Saying: 'This May be True in Theory, but it does not Apply in Practice'," in *Kant's Political Writings*, ed. Hans Reiss, trans. H. B. Nisbet (Cambridge, UK: Cambridge University Press, 1970), 62.

4. For Foucault's use of "*panopticon*" see his "The Eye of Power," in *Power/Knowledge: Selected Interviews and Other Writings 1972–1977*, ed. Colin Gordon, trans. Gordon et al., (New York: Pantheon Books, 1980), 146–165; and *Discipline and Punish: The Birth of the Prison* (New York: Random House, 1978), 195–228.

5. Maurice Merleau-Ponty, *In Praise of Philosophy*, trans. with a Preface by John Wild and James M. Edie (Evanston, IL: Northwestern University Press, 1963), 30. I invoked Merleau-Ponty's lecture in my own inaugural lecture at the University of Notre Dame on March 21, 1980. See my "Political Philosophy Today," in *Polis and Praxis: Exercises in Contemporary Political Theory* (Cambridge, MA: MIT Press, 1984), 15–46.

6. Merleau-Ponty, *In Praise of Philosophy*, 34–36.

7. Hans-Georg Gadamer, *Praise of Theory: Speeches and Essays*, trans. Chris Dawson (New Haven: Yale University Press, 1998), 31, 35.

8. Gadamer, *Praise of Theory*, 19, 22–24, 29. As one may recall, Gadamer pointedly called his chef d'oeuvre "Truth and Method" with an explicit primacy of the former over the latter. See *Truth and Method*, 2nd rev. ed., trans. Joel Weinsheimer and Donald G. Marshall (New York: Continuum, 1989).

9. See "In Praise of Philosophy," in Jacques Derrida, *Eyes of the University: Right to Philosophy 2*, trans. Jan Plug and Others (Stanford, CA: Stanford University Press, 2004), 157, 159. Already two years before this interview, in 1978, Derrida had reflected on "The Crisis in the Teaching of Philosophy"; see his *Who's Afraid of Philosophy? Right to Philosophy 1*, trans. Jan Plug (Stanford, CA: Stanford University Press, 2002), 99–116.

10. "In Praise of Philosophy," 160–163. The proposed International College of Philosophy was in fact established in 1982; see "Titles (for the Collège International de Philosophie) (1982)," in the same volume, 195–215.

11. See Derrida, "The Right to Philosophy from a Cosmopolitical Point of View," in *Ethics, Institutions, and the Right to Philosophy*, trans. and ed. Peter Pericles Trifenas (Lanham, MD: Rowman & Littlefield, 2002), 5–7, 9–10. For Kant's text see "Idea for a Universal History with a Cosmopolitan Purpose" (1784), in *Kant's Political Writings*, ed. Hans Reiss, trans. H. B. Nisbet (Cambridge, UK: Cambridge University Press, 1970), 41–53.

12. Derrida, "The Right to Philosophy," 10–13, 15. Among the dangers he mentions are especially "techno-economic, indeed scientific-military imperatives" and a "techno-economic-military positivism." In light of these dangers, the right to philosophize "becomes increasingly urgent, irreducible" (15). Derrida in this context links the cosmopolitan right to philosophize with one of his signature ideas: a "democracy to come" (13–14). On this idea see my "Jacques Derrida's Legacy: 'Democracy to Come'," in *The Promise of Democracy: Political Agency and Transformation* (Albany, NY: State University of New York Press, 2010), 117–134.

13. See Martin Heidegger, "Modern Natural Science and Technology: Greetings," in *Research in Phenomenology*, vol. 7 (1977), 3; also Heidegger, *Zollikoner Seminare: Protokolle-Gespräche-Briefe*, ed. Medard Boss (Frankfurt-Main: Klostermann, 1988), 157 (November 23, 1965). Compare also Heidegger, *Being and Time*, trans. John Macquarrie and Edward Robinson (New York: Harper & Row, 1962), 21–35.

14. Heidegger, *Einleitung in die Philosophie*, ed. Otto Saame and Ina Saame-Speidel (*Gesamtausgabe*, vol. 27; Frankfurt-Main: Klostermann, 1996), 2–3.

15. Heidegger, *Einleitung in die Philosophie*, 4–5, 11. The statement that *Dasein* cannot "possess itself" (*ganz und gar nicht das Eigentum seiner selbst ist*) implies a radical rejection of Max Stirner's *Der Einzige und sein Eigentum* (1845); English: *The Ego and His Own*, trans. Steven T. Byington, ed. John Carroll (New York: Harper & Row, 1971).

16. Heidegger, *Einleitung in die Philosophie*, 22–23.

17. Heidegger, *Beiträge zur Philosophie (Vom Ereignis)*, ed. Friedrich-Wilhelm von Herrmann (*Gesamtausgabe*, vol. 65; Frankfurt-Main: Klostermann, 1989), 37–39. For an English translation (not entirely followed above) see Heidegger, *Contributions to Philosophy (of the Event)*, trans. Richard Rojcewicz and Daniela Vallega-Neu (Bloomington, IN: Indiana University Press, 2012), 31–33.

18. Heidegger, *Besinnung*, ed. Friedrich-Wilhelm von Herrmann (*Gesamtausgabe*, vol. 66; Frankfurt-Main: Klostermann, 1997), 16–17. For an English translation (not entirely followed above) see Heidegger, *Mindfulness*, trans. Parvis Emad and Thomas Kalary (London: Continuum, 2006), 12–14. See also my "Heidegger on *Macht* and *Machenschaft*," *Continental Philosophy Review*, vol. 34 (2001), 247–267; and "The Underside of Modernity: Adorno, Heidegger and Dussel," *Constellations*, vol. 11 (2004), 102–120.

19. Heidegger, *Besinnung*, 15, 207–209; *Mindfulness*, 11, 183–185.

20. Heidegger, *Was heisst Denken?* (3rd ed.; Tübingen: Niemeyer, 1971), 1–3. For an English translation (not always followed above) see Heidegger, *What is Called Thinking?* trans. J. Glenn Gray (New York: Harper & Row, 1968), 3–5.

21. *Was heisst Denken?* 4–6; *What is called Thinking?* 8–10. In an earlier context, Heidegger had stated provocatively: "Science wants to know nothing about 'nothing'." See "What is Metaphysics?" in *Martin Heidegger: Basic Writings*, ed. David F. Krell (New York: Harper & Row, 1977), 98.

22. *Was heisst Denken?* 17–18, 94; *What is Called Thinking?* 32, 145.

23. Heidegger, "Was heisst Denken?" in *Vorträge und Aufsätze*, Part 2 (3rd ed., Pfullingen: Neske, 1967), 8, 14–15.

24. Heidegger, "Wissenschaft und Besinnung," in *Vorträge und Aufsätze*, Part 1 (3rd ed., Pfullingen: Neske, 1967), 49–50, 55, 62. To round out this review, mention should also be made of two later lectures delivered in 1955: "Was ist das—die Philosphie?" and "Gelassenheit." For the former see Heidegger, *What is Philosophy?* trans. Jean T. Wilde and William Kluback (New Haven, CT: College and University Press, n.d.) where he distinguishes again between calculating and mindful thinking. For the second see *Gelassenheit* (Pfullingen: Neske, 1959), translated as *Discourse on Thinking*, trans. John M. Anderson and E. Hans Freund (New York: Harper & Row, 1966). In the latter translation, the phrase *"besinnliches Denken"* is misleadingly labeled "meditative thinking."

25. Heidegger, *Was heisst Denken?* 12, 26; *What is Called Thinking?* 38, 46. The citation is from Friedrich Nietzsche, "Thus Spoke Zarathustra," Part IV, in *The Portable Nietzsche*, ed. Walter Kaufmann (New York: Viking Press, 1968), 417, 421.

26. Heidegger, *Gelassenheit*, 20; *Discourse on Thinking*, 52. The dangers of a purely technical and calculating "theorizing" for humanity have been even more forcefully castigated by Maurice Merleau-Ponty in these statements (penned shortly before his death): "Thinking 'operationally' has become a sort of absolute artificialism, such as we see in the ideology of cybernetics, where human creations are derived from a natural information process, itself conceived on the model of machines. If this kind of thinking were to extend its reign to man and history; if . . . it were set out to construct man and history on the basis of abstract indices . . . then, since man really becomes a *manipulandum* he takes himself to be, we enter into a cultural regimen where there is neither truth nor falsity concerning man and history, into a sleep or a nightmare from which there is no awakening." The remedy proposed by Merleau-Ponty is a return to embodied human existence: "Scientific thinking which looks on from above and thinks of the object-in-general, must return to the 'there is' which underlies it; to the site, the soil of the sensible and opened world such as it is in our life and for our body." See his "Eye and Mind," in *The Primacy of Perception and Other Essays*, ed. James M. Édie, trans. Carleton Dallery (Evanston, IL: Northwestern University Press, 1964), 160.

27. Some evidence of "Western bias" can be found in Heidegger's lecture "What is Philosophy" of 1955. However, this bias needs to be juxtaposed to his appreciation of Asian philosophy and his great resonance in non-Western, especially Japanese contexts. See, for example, Graham Parkes, ed., *Heidegger and Asian Thought* (Honolulu, HI: University of Hawaii Press, 1987); Hartmut Buchner, ed., *Japan und Heidegger: Gedenkschrift zum 100. Geburtstag* (Sigmaringen: Thorbecke Verlag, 1989); and Florian Vetsch, *Martin Heidegger's Angang der interkulturellen Auseinandersetzung* (Würzburg: Königshausen and Neumann, 1992). Regarding thinking and judging compare Ronald Beiner, *Political Judgment* (Chicago: University of Chicago Press, 1983); Hannah Arendt, *Lectures on Kant's Political Philosophy* (Chicago: University of Chicago Press, 1992); Leslie Paul Thiele, *The Heart of Judgment: Practical Wisdom, Neuroscience, and Narrative* (New York: Cambridge University Press, 2006).

28. Heidegger, "Letter on Humanism," in *Martin Heidegger: Basic Writings*, ed. David F. Krell (New York: Harper & Row, 1977), 241–242.

TWO

Mindful Praxis

Beyond "Action Theory"

In action through non-action, there is nothing that is not properly done.
—*Daodejing*, Chapter 3

The distinction between calculation and mindfulness, elaborated in the preceding chapter, is relevant not only for "cognitive" processes, but for human life as a whole, including prominently the life of practice or practical conduct. In this respect, it is sometimes assumed that mindful thinking coincides with abstract meditation or speculation—an assumption which flies in the face of the world-openness of thinking, its attentiveness to everything that "calls for" thought. In other words, mindfulness is not the label for a "philosophy of mind" where mind only "minds" itself. It is precisely its world-openness, however, which exposes thinking to the opposite danger: that of serving as a mere decoy for world-appropriation and control.

This danger is by no means fortuitous. It so happens that the most prominent and influential theory of practice in contemporary social science is the theory of "rational action" or "rational choice," a doctrine according to which human reason—or else self-interest—stipulates the goal of an action and also the most efficient means for reaching that goal. From this angle, an action is said to be most efficient if it yields maximal benefit or profit for minimal effort. As can be seen, human beings here are nothing but calculating machines constantly tabulating winnings and losses, with the latter having a purely negative value.

From the perspective of mindfulness, human practice or conduct has a completely different—non-appropriating or non-predatory—quality. A good initial glimpse of this quality is provided by the Aeschylean phrase "*pathei mathos,*" meaning "taught through suffering" or "having learned

29

the hard way." The phrase indicates that suffering is not always a negative loss, but can have an important seasoning or humanizing quality.
Commenting on the phrase, Hans-Georg Gadamer thoughtfully observes
that its point is not only that "we become wise through suffering," rather,
it means something more radical: namely, that we are not in control of
our experience or of what "discloses itself" in experience; that, precisely
due to our finitude, we have to eschew absolute cognition and remain
open to ever new experiences.[1] Thus, what Aeschylos's phrase demands
of us is not be satisfied with a finished system of ideas or principles
which then might be imposed on human conduct in order to permit a
rational calculation of winnings and losses. The demand, at the same
time, does not endorse a blind "activism" which often is only a form of
self-promotion or self-indulgence. Rather, the task is to calibrate thinking
and practice or practical experience in a mindful way—a way which was
intimated in Hannah Arendt's statement that her effort was to grasp
thoughtfully "what is underfoot" or simply "to think that we are doing."[2]

   In the following, I want to discuss three major perspectives which
illustrate in exemplary fashion the meaning of mindful praxis: the philosophy of Martin Heidegger; the religious thinking of Raimon Panikkar;
and the Asian notion of *wu-wei*. In all three cases, the focus will be on the
correlation of understanding and doing, mindfulness and praxis.

HEIDEGGER: LETTING BE

The issue of a mindful praxis is beset with particular difficulties and
pitfalls in the case of Heidegger. The difficulties derive from two widespread biases or misunderstandings which, in their combination, make
access to the issue impossible. On the one hand, there is the view of
Heidegger as a recondite metaphysical author whose writings exceed the
bounds of ordinary intelligibility. Seen in this way, his status is that of a
"seer" or *guru* far removed from worldly affairs and practical concerns.
On the other hand, there is the recollection of his political engagement in
1933—an engagement which, in the view of many, reduces his entire
philosophy to a mere appendage of a deplorable practice. Clearly, both
ideas erect powerful obstacles to the inquiry attempted here; but they are
not insuperable. The image of a seer or *guru*, in my understanding, ac-
cords ill or not at all with an opus offering rigorous analyses of difficult
thinkers like Leibniz, Kant, Hegel, and others. On the other hand, the
episode of 1933—although deplorable—could (or should) be seen not so
much as an end-point but rather as a painful learning experience (*pathei
mathos*) which in many ways fueled Heidegger's famous "*Kehre*," his
turning away from modern anthropocentrism and the centrality of the
Cartesian *cogito*. To a large extent, it was precisely this "*Kehre*"—whose

seeds, to be sure, reach farther back in time—which enabled him to tackle the issues of thinking and praxis in a novel and unconventional way.[3]

Broadly stated, the novelty of Heidegger's approach consists in his resolute exit from the theory-practice binary which has dominated Western philosophy since the onset of modernity. In terms of this binary, theory means the endeavor to obtain a detached grasp, an "objective" spectatatorial overview of the world and all its phenomena. On the other hand, practice or action denotes the pursuit of a "subjective" project or agenda with the aim of realizing some advantage or benefit for the agent. By adopting as primary and prior issue the "question of Being"—what it means for anything to "be"—Heidegger undercut the modern bifurcation. This undercutting effort is nowhere more clearly stated than in the "Letter on Humanism" (1946) where Heidegger explicitly asks how theory and practice are related to mindfulness and the "question of Being." With regard to the first term of the binary, the answer is crisp: Thinking "exceeds all pure contemplation or theoretical observation because it cares first of all about the light or clearing in which seeing as '*theoria*' can live and move" (or becomes possible). Does this mean then that thinking is a practice? Yes, Heidegger answers, but a practice which "surpasses all action (in the sense of fabrication or production)." What needs to be realized is that mindful thinking "towers above such action or production not through the grandeur of its achievements nor through its effected consequences, but through the poverty of its inconsequential accomplishment." This poverty or simplicity alienates a success-oriented age. For, "we tend to conceive thinking after the model of scientific knowledge and research projects. And we measure action in terms of impressive and successful results."[4]

If genuine action is neither pure theory (or derived from theory) nor a mode of fabrication, what in Heidegger's account is the meaning of mindful praxis, that is, a praxis attentive to the question of Being? Is there an action which is not the handmaiden of cognitive formulas nor subservient to chosen ends beyond itself? The opening of the "Letter on Humanism" provides an answer to these queries (though an answer which has been too widely ignored). As Heidegger states: "We are still far from pondering the essence of action decisively enough. We view action only as causing an effect, whose effectiveness is defined in terms of its utility." What is completely missed in this view is the integral quality of action, the fact that praxis carries its worthiness or its "being" within itself (as is evident in the case of flute-playing so often mentioned by Aristotle). In Heidegger's lapidary formulation: "But the essence of action is fulfillment (*Vollbringen*), where 'fulfilling' means to unfold something into the fullness of its being, or to lead it forth into its fullness (*producere*)." This statement does not stand by itself or in a vacuum; rather, it is intimately connected with Heidegger's idea of thinking as mindfulness, that is, as openness to what "calls upon" thinking—which is the question of Being.

As he adds: "Only what already and genuinely 'is' can really be fulfilled. But what 'is' above all is Being." Thus, thinking as mindfulness means an openness to the question of Being, a readiness to respond to its call and let it "be" as a basic summons. In the same way, mindful praxis means an action which reaches its fulfillment by being attentive to the call and pursuing the path (of truth and freedom) to which it summons.[5]

What is important to remember here is that action, in the sense of a mindful praxis, is not a solitary or solipsistic conduct, but remains embedded in a comprehensive fabric of experience, a fabric which includes both other human beings and the phenomenon of nature. With respect to the latter, praxis is guided by mindfulness in the sense of ecological awareness and responsiveness; it thus eschews efforts to dominate or exploit nature in the pursuit of narrowly anthropocentric goals. The same attitude pervades interhuman relations, that is, the domain of human "co-being" or inter-being (*Mitsein*). Far from seeking to control, manipulate or manage other human lives, mindful praxis respects and cherishes the integral quality of fellow-beings, that is, nurtures and assists their own possibilities for being. What assumes center stage here is Heidegger's key notion of "letting be" (*Seinlassen*)—a notion which is far removed from negligence or indifference. On the contrary, precisely by letting fellow-beings be, mindful praxis cares deeply about and for the other—although "caring," in this and other relevant instances, has no truck with the promotion of self-interest. In many ways, "letting be" can be seen as a guiding motif in all of Heidegger's writings, including writings antedating the so-called "*Kehre*." A good example is the lecture course he offered in 1928/29 under the title "Introduction to Philosophy" (*Einleitung in die Philosophie*). In this lecture course—which was published in 1996—Heidegger presented "letting be" not just as one possible kind of praxis but as the "primordial praxis" (*Urhandlung*) of human *Dasein*.

As on other occasions, the lecture course distinguishes genuine praxis clearly from sheer activism and especially from modes of instrumental action (*poiesis*) seeking to produce a specific outcome or effect. As Heidegger points out, the distinction is difficult today given the prevalent opinion "that 'action' and 'doing' occur only where life is hectic, where business booms, and power and mastery hold sway." Against this opinion, the lecture course marshals the Greek insight that praxis does not aim at ulterior effects, but rather carries its goal and fulfillment (*Vollbringen*) in itself—with the result that it can be termed autonomous and self-fulfilling (*autarkes* and *autoteles*). Moreover, as Aristotle had continuously emphasized, praxis is a doing which not only carries its end or fulfillment in itself, but one which completes or "fulfills the agent or actor." The latter achievement, in Heidegger's view, is possible because human *Dasein* is constituted by the "care" for Being and basically oriented toward the meaning and truth of Being. Any action, therefore, which lets itself be guided by this truth has a transformative and deeply humanizing (a fulfilling) quality. In terms of the lecture course, "letting-be is a praxis (*Tun*) of the highest and most primordial sort, and possible only on the basis of the inner core of our existence: its freedom" (toward the truth of Being). Differently phrased: Letting be is a doing "that allows beings to become manifest (in the truth). This is the primordial praxis (*Urhandlung*)."[6]

What one should note here is the peculiar in-between character of the "doing" involved in praxis. Doing is not a linear goal-directed "project" under the agent's firm control; nor is it a destiny or fate befalling the agent from the outside. In a sense, one may say that praxis occurs in the "middle voice," a voice combining action and passivity, doing and experiencing or underdoing (*pathei mathos*). In Heidegger's words, praxis seems to proceed from "a spontaneity located entirely in ourselves"; however, by encountering and letting things and beings "be," it displays "a peculiar passivity or receptivity." Such receptivity implies openness and a kind of "self-surrender" (*Sichfreigeben*) in favor of beings "so that they can show themselves how they are." In a slightly different idiom, praxis as letting-be occurs between or beyond interest and disinterest. Seen as a doing, praxis allows beings or things to be themselves in their truth; in this "allowing" we are intensely involved or engaged because we are constituted by "care" (for Being). However, care here is free from narrow self-interest or the desire for individual gain. As Heidegger points out, in letting-be we adopt a stance of "equanimity" (*Gleichgültig*) which might also be called "releasement" (*Gelassenheit*) and which is *not* indifference. In terms of the lecture course: "Such equanimity is possible only in the modality of care (*Sorge*). The letting involved in this allowing, however, is not a bare omission (*Unterlassen*)." With regard to human co-being or inter-being, letting-be inaugurates a community of free beings equally oriented toward truth (and far removed from both selfish libertarianism and coercive communalism).[7]

PANIKKAR: CO-CREATION

In the context of recent Western philosophy, Heidegger's delineation of mindful praxis as releasement is unparalleled in its subtlety and depth of insight. However, pointers in a similar direction can also be found in the writings of other thinkers, especially those influenced by Heidegger's work. A good example is Spanish-Indian religious philosopher Raimon Panikkar. The *magnum opus* of his later years reveals the influence already in its chose title: The Rhythm of Being. As in Heidegger's case, the term "Being" for Panikkar does not denote an abstract concept or a fixed substance but rather the enabling ground (and un-ground) of all possible beings, a ground which is forever in a process of renovation and creative transformation. As he writes, the word Being in his text designates "the overall symbol that encompasses, in one way or another, all to which we

may meaningfully say 'is,' anything that enters the field of our awareness." What needs to be remembered is that Being is "not a thing or
entity," but the ground which enables things to be. Moreover, "rhythm"
is not something added to or superimposed on Being: "There is nothing
'outside' Being." Hence, the rhythm of Being is simply "Being's rhythm";
just as Being "is *in* all beings, the rhythm belongs to Being itself." Caught
up in this rhythm, human beings are not isolated or fully autonomous
agents nor passive victims of an inexorable fate; rather, they are participants in an ongoing happening or disclosure—although genuine participation requires careful attentiveness, attunement, and self-transforma-
tion or self-trancendence.[8]

As in Heidegger's case, attentiveness or attunement for Panikkar requires a lack of blinders, that is, an openness of thought to what "calls
for" thinking. In a vintage Heideggerian idiom, he links thinking with
Being in a close "yin-yang" connection. In his words: 'Thinking thinks
Being; Being begets thinking. One might even risk saying: Being 'beings
thinking.'" This means that, in a sense, thinking is "an activity of Being,"
that is, a responsiveness to the latter's call. For this responsiveness to be
possible, a certain "purification" has to happen—which is not only a
cleansing of the mind but a "turning" of the entire human being.
Wherever such a cleansing or turning happens, one can speak of "releasement" (*Gelassenheit*) or the ability of "letting be." In order to approach
this state, Panikkar observes, it is important "not to put up obstacles for
the spirit, or barriers to divine grace, letting the *Tao* be, becoming trans-
parent, and renouncing the fruits of action." In more traditional Christian
language, turning is also captured in the notion of *"metanoia"* —with Panikkar commenting that *metanoia* "means more than a 'change of mind'; it
means to overcome the [purely] mental [or rational]." In a different
Christian vocabulary, one can speak here of human participation in the
ongoing disclosure of the divine, that is, in what St. Bonaventure called
the *"creatio continua"* where humans are seen as co-creators in the unfolding of salvation history.[9]

Viewed as a disclosure of Being or a mode of participation, thinking is
closely related to, and even coterminous with, doing or praxis. As Panikkar notes—again in a very Heideggerian vein – thinking should not be
reduced to the work of reason, intellect or "mind" (all variations of the
Cartesian *cogito*). This "sad reductionism," he states, has led to "unnatu-
ral quietism and lopsided idealisms"; above all, it has led to the "degrad-
ing of the importance of 'doing,' as if spiritual life could be reduced to
'theory'" (in the sense of pure cognition). Referring to the beginning of
John's gospel, he acknowledges the invocation of *"logos"* —a term frequently translated as "word." As he cautions, however, word here is not
merely a mental construct; rather, a word is "also sound and action."
More pointedly stated: "A word is intrinsically preformative; a real word
does, it is *praxis* and *poiesis*." At this point, an important distinction needs

to be made: that between doing and (technical) making. The confusion of these terms has had disorienting consequences. Thus, in theology, we encounter the "anthromorphic and ultimately wrong idea of a divine Maker, architect or worse engineer," that is, of God as a super-technician who can make and unmake his products at will. Fashioned in the image of this Maker, human beings also are frequently construed as tool-makers (*homo faber*). In opposition to this view, Panikkar presents human beings not as engineers but as care-takers and custodians, assigning to them the role not of "absolute kings of creation" but as its "gardeners" (reminiscent of Heidegger's notion of "shepherds"). What is needed to grasp this view is an exit from anthroprocentrism and a move toward an "alternative anthropology" which considers human beings as "a microcosm, and eventually even a 'micro-theos.'"[10]

As in Heidegger's case again, human praxis for Panikkar occupies an in-between space between willful constructivism and fatalism. "The task of transforming the world," he states, "is not achieved by a merely passive attitude nor by sheer activism. It is brought about by being co-operators (*synergoi*) with the divine" in its ongoing disclosure. More firmly than the German philosopher, Panikkar at this point is willing to enter the domain of politics and public engagement. As he writes: "The world cries out for a radical change that cannot be merely theoretical, without a grounding in praxis"—a praxis which, in turn, needs to be informed by thinking. "Who or what," he adds, "will put a halt to the lethal course of technocracy? More concretely: Who will control armaments, polluting industries, cancerous consumerism, and the like? And who will put an end to the unbridled tyranny of money?" Momentous challenges of this kind demand, in his view, an equally momentous or unprecedented recourse to mindful praxis, that is, a political praxis infused with radical mindfulness—where the latter also includes an openness to religious teachings. Contrary to the vogue of fundamentalism in many parts of the world, however, Panikkar strongly counsels against an indiscriminate merger of politics and religion—where one side dominates or instrumentalizes the other. As he says: Where institutional religion dominates, humanity "suffocates under an unbreathable heaven on earth"; on the other hand, where pure power politics holds sway, human life withers "from the lack of oxygen from heaven." What is needed, hence, is a careful calibration of mundane and spiritual concerns, of "immanent" and "transcendent" dimensions of life—a calibration which is precisely the hallmark of the "non-dualistic" or "cosmotheandric" outlook developed in Panikkar's writings.[11]

KARMA YOGA AND WU-WEI

In the *Rhythm of Being*, Panikkar refers repeatedly to classical Indian teachings, especially the teachings of the Upanishads and the *Bhagavad Gita*. As he reminds us, the key notion of the Upanishads is *"brahman"* — which is by no means an abstract concept to be grasped cognitively by the mind (or *cogito*). One of the most famous utterances of the Upanishads is *"aham-brahmasmi"* that is, "I am *brahman*." However, as Panikkar points out, this utterance—one of the "great sayings" (*mahavakya*) of ancient India—does not mean that "my ego is *brahman*," but rather that *"brahman* is 'I am'" or the ground of Being. When rid of all selfish individuality, we are able to experience "I am" in such a way that our own being merges with Being and Becoming (or the "rhythm of Being"). Differently put: *brahman* is not an "object of thought," but rather a premise of thought and an emblem of transformative mindfulness. In the words of another Upanishad, *brahman* is that "which is not minded by the mind, but by which, they say, the mind is minded." What statements of this kind bring into view is a fundamental holism or non-dualism which is the heart of classical Vedanta or *Advaita* Vedanta. The gist of this outlook is captured in another Upanishad which states boldly: "You cannot hear the hearer of hearing; you cannot think the thinker of thinking; you cannot know the knower of knowing."[12]

The insight of the Upanishads is carried forward in the *Bhagavad Gita*, but with a greater accent on mindful praxis—grasped under the heading of *"yoga."* The text distinguishes between different kinds of *yoga*, including the praxis of wisdom (*jñana*) and devotion (*bhakti*); however, the main emphasis is on the *yoga* of action, termed *"karma yoga."* Paralleling or anticipating the teachings of Heidegger and Panikkar, the basic character of *karma yoga* resides in its dismissal of egocentrism, its abandonment of the pursuit of self-serving aims. In the words of the *Gita*: "Set your heart upon action or work (*karma*), never on its reward. Thus, work not for the sake of a personal benefit; yet never cease to perform your work." As can be seen, *karma yoga* occupies a mid-point between action and inaction, between doing and not doing: it is a non-possessive and non-coercive action released from attachment to the fruits of action; at the same time, it is an action done in the service of others and of "world-maintenance" (*loka-samgraha*). Released from possessive attachments, action can be called "pure" or also "consecrated"—which is precisely the language favored by the *Gita*. As we read there: "Great is the person who, free from attachments, and with a mind ruling its powers in harmony, works on the path of *karma yoga*, the path of consecrated action. . . . The world is in the bonds of causation, unless the action is consecration." The distinction between genuine *karma yoga* and selfish action is for the *Gita* also the difference between wisdom and foolishness, of *jñana* and *avidya*: "Even as the unwise work selfishly in the bondage of selfish works, let the wise man work unselfishly for the good of all the world."[13]

The idea of unselfish conduct as a mid-point between action and non-action can also—and prominently—be found in East Asian thought, especially in the teachings of Laozi and Zhuangzi. Everyone is familiar with the opening lines of the *Daodejing* (ascribed to Laozi), lines which speak of a mindfulness beyond epistemic cognition: "The way (*tao*) that can be spoken of is not the real way; the name that can be named is not the real/eternal name. Nameless was the beginning of heaven and earth." These lines are almost instantly followed by a passage which—in words reminiscent of the *Gita*—cautions against attachments clinging to the fruits of desire: "Free from desire, you realize the mystery; caught in the web of desires, you only grasp surface manifestations." Freedom from attachments requires releasement from the burden of things and appearances, especially the appearance of a stable and meritorious selfhood. In terms of a later passage: "Standing tiptoe a man loses his balance; rushing ahead he misses his pace; trying to shine (or outshine others), he dims his own light . . . If you want to accord with the way (*tao*), just do your work, then let go." This advice to "let go" is supplemented by repeated warnings against selfish conceit and intellectual pretenses: "Leave off fine learning. End the nuisance of saying yes to this and perhaps to that—distinctions with little difference, categorical this, categorical that." And at another point we read: "Empty your mind of all conceit; let your heart be fully at peace . . . When you realize where you come from, you naturally become tolerant, open-minded and open-hearted, kind-hearted as a grandmother, dignified as a king."[14]

As repeatedly stated before, a crucial corollary of non-epistemic thinking is mindful praxis which, in the Daoist tradition, is captured in the expression of "*wu-wei*," often translated as "inactive action" or "non-assertive action." Examples or illustrations of this notion are frequent in the *Daodejing*. Thus, according to an early chapter: "The sage (or sane) person acts without doing anything, and teaches without inculcating anything. As things arise, he/she lets them come; as things disappear, he/she lets them go. He/she has, but does not possess, acts but does not appropriate." And at a later point we read, with distinct political overtones: "A realm is governed by ordinary acts, but the world is governed by no [controlling] acts of all. . . . Therefore, the sane person says: let go of laws and the people become honest; let go of economics and the people become prosperous; let go of preaching and the people become decent."[15] The emphasis on *wu-wei* is carried forward resolutely in the *Zhuangzi*. Thus, we read in one of the so-called "inner chapters" of this text: "Those of old who promoted the way (*tao*) stressed tranquillity to foster understanding. They understood the way, but did not employ this understanding to promote intrusive action (*wei*) . . . When understanding and retinence are both cultivated, harmony emerges from our nature [or without prodding]." In line with this outlook, the chapter expresses suspicion of anyone who wants to "take charge of the world (*wei tianxia*)," that is, "to manage or control everything under the sky." The title of another chapter says explicitly "Let it Be, Leave it Alone."[16]

Careless readers are sometimes tempted to compare the Chinese *wu-wei* with "*laissez-faire*," the modern doctrine of market liberalism guided by a "hidden hand"; however, the comparison is completely mistaken—for a simple reason. The modern doctrine is founded on acquisitive individualism, that is, on individual agents seen as "utility maximizers" seeking to obtain the greatest benefit for the least effort. This conception runs counter to the Chinese notion of effortless conduct, of a non-assertive and non-acquisitive action devoid of self-centered aims. According to a famous passage in the *Daodejing*, *wu-wei* is predicated on the agent's self-abandonment or self-forgetting: "Because the sane person does not grab hold of anything, he/she does not lose anything." This passage is paralleled by the equally famous statement in the *Zhuangzi*: "The genuine person (*zhenren*) is beyond self; the spiritual person is beyond merit; the sage person is beyond reputation [or name]." The same text has this to say about the genuine or authentic person: "The *zhenren* of old did not lord over minorities, did not show off achievements, and did not scheme over affairs. Such a person could transgress without being overbearing, or hit the mark without being self-satisfied or boastful." And another passages adds: "The *zhenren* is unified with Heaven and connected with the way (*dao*); holding fast to the one, he/she nourishes and fosters the myriad kinds. Keeping the heart-and-mind of Heaven in the bosom, he/she generates strength (*de*) and uses *wu-wei* to envelop aspirations, worries, thoughts, and intentions."[17]

Much has been written about *wu-wei*—not always with sufficient discernment. In my view, its main contemporary significance resides precisely in its sharp contrast to the liberal or libertarian idea of "*laissez-faire*"—a doctrine which, far from generating wholeness and peace, is the source of class warfare and even of global strife (at least when rigidly applied). The proper meaning of *wu-wei* has been correctly delineated by China-scholar Roger Ames in several of his writings. As he notes, the term is frequently understood as a synonym for "passivity, femininity, quietism, pure spirituality," an outlook favored by "artists, recluses, and religious mystics"; this reading, however, falls short by simply equating the term with inaction. What is missed is the "in-between" character of *wu-wei*, its functioning in the "middle voice"—where that term implies a radical relationality or wholeness. Viewed in this light, *wu-wei* can be seen to imply a "productively creative relatedness"; free of unilateral assertiveness, relational or responsive action/reaction furnishes a balanced space for "novel possibilities of a richer creativity."[18]

Going beyond historical exegesis, another writer has linked *wu-wei* with John Dewey's conception of modern democracy seen as a cooperative framework of action and interaction in an ongoing process of creative innovation. In his words: Both Daoist philosophers and Dewey appeal to "the natural cooperation and 'cooperative intelligence' of people as the only means or 'way' of attaining fulfillment in human life and of addressing and resolving adequately social and political problems." As the same writer adds, following both Dewey and Chinese teachings, the interplay of the "vibrant community of cooperative inquirers" may well be the best means of confronting the mounting crises of our age—crises which threaten to throw the world into chaos and destruction.[19]

NOTES

1. Hans-Georg Gadamer, *Truth and Method*, 2nd rev. ed., trans. Joel Weinsheimer and Donald G. Marshall (New York: Crossroad, 1989), 356–357.

2. Hannah Arendt, *The Human Condition: A Study of Central Dilemmas Facing Modern Man* (Chicago: University of Chicago Press, 1958), 4. Compare also Arendt, *Between Past and Future: Six Exercises in Political Thought* (Cleveland/New York: Meridian Books, 1965), where she writes (14): "My assumption is that thought itself arises out of incidents of living experience and must remain bound to them as the only guideposts by which to take its bearings." See also my "Action in the Public Realm: Arendt between Past and Future," in *The Promise of Democracy: Political Agency and Transformation* (Albany, NY: State University of New York Press, 2010), 83–97.

3. For some background see my *The Other Heidegger* (Ithaca, NY: Cornell University Press, 1993); "Resisting Totalizing Uniformity: Martin Heideggar on *Macht* and *Machenschaft*," in *Achieving our World: Toward a Global and Plural Democracy* (Lanham, MD: Rowman and Littlefield 2001), 189–209, and "The Underside of Modernity: Adorno, Heidegger, and Dussel," *Constellations*, vol. 11 (2004), 102–120.

4. Heidegger, "Letter on Humanism," in *Martin Heidegger: Basic Writings*, ed. David F. Krell (New York: Harper & Row, 1977) 239–240.

5. "Letter on Humanism," 193. For the relation of the question of Being and truth see Heidegger, "On the Essence of Truth" ("*Vom Wesen der Wahrheit*") in *Martin Heidegger: Basic Writings*, 117–141.

6. Heidegger, *Einleitung in die Philosophie*, ed, Otto Saame and Ina Saame-Speidel (*Gesamtausgabe*, vol. 27; Frankfurt-Main: Klostermann, 1996), 102–1–3, 173–178, 183–184.

7. Heidegger, *Einleitung in die Philosophie*, 74–75, 102–103. For a more detailed discussion see my "Agency and Letting-Be: Heidegger on Primordial Praxis," in *The Promise of Democracy*, 67–81.

8. Raimon Panikkar, *The Rhythm of Being: The Gifford Lectures* (Maryknoll, NY: Orbis Books, 2010), 51–52.

9. Panikkar, *The Rhythm of Being*, xxviii, 2, 32, 35. As Panikkar adds (35): "Only when the heart is pure are we in harmony with the 'real,' in tune with reality, able to hear its voice, and truly speak its truth, having become adequate to the movement, the Rhythm of Being."

10. Panikkar, *The Rhythm of Being*, 348–349, 351. The distinction between making and doing is aptly captured in Hannah Arendt's differentiation between "work" and "action." See her *The Human Condition*, 119–223. For Heidegger's notion of *Dasein* as "shepherd of Being" see his "Letter on Humanism," in *Martin Heidegger: Basic Writings*, ed. David F. Farrell (New York, Harper & Row, 1977), 210.

11. Panikkar, *The Rhythm of Being*, 350, 356–358. Compare in this context my "Postsecular Faith: Toward a Religion of Service," and "Religion and the World: The Quest for Justice and Peace," in *Integral Pluralism* (Lexington, KY: University of Kentucky Press, 2010), 67–83, 85–101.

12. Panikkar, *The Rhythm of Being*, 100, 184, 202. The references are to Kena Upanishad (I, 6) and Brihadaranyaka Upanishad (III, 4, 2).

13. See *The Bhagavad Gita*, trans. Juan Mascaró (London: Penguin Books, 1962), 52 (Chapter 2, verse 47); 56–57 (Chapter 2, verses 7, 9); 58 (Chapter 3, verse 25).

14. The citations are from Chapters 1, 16, and 20 of the *Daodejing*. I have relied on three different English translations: Stephen Mitchell, *Tao Te Ching: A New English Version* (New York: Harper Perennial, 1991); Witter Bynner, *The Way of Life: According to Lao Tzu* (New York: Perigee Books, 1972); *Lao Tzu: Tao Te Ching*, trans. D.C. Lau (London: Penguin Books, 1963).

15. See *Daodejing*, Chapters 2 and 57.

16. The references are to Chapters 11 and 16 of the *Zhuangzi*. I have used these texts: Burton Watson, *The Complete Works of Chuang Tzu* (New York: Columbia University Press, 1968); and A. C. Graham, *Chuang-Tzu: The Inner Chapters* (London: Unwin, 1989).

17. See *Daodejing*, Chapter 64; *Zhuangzi*, Chapters 1 and 6. See also Daniel Coyle, "On the *Zehnren*," in Roger T. Ames, ed., *Wandering at Ease in the Zhuangzi* (Albany, NY: State University of New York Press, 1998), 197–210; and Chris Jochim, "Just Say No to 'No Self' in *Zhuangzi*," 35–74.

18. See Ames, "Introduction," in *Wandering at Ease in the Zhuangzi*, 5, 7–8. Compare also Ames, *The Art of Rulership: A Study of Ancient Chinese Political Thought* (Albany, NY: State University of New York Press, 1994), Chapter 2, "Wu-Wei," 28–64. In Ames' view, there is not a deep gulf between Daoism and Confucianism, at least with regard to *wu-wei* (an interpretation I share).

19. See Henry G. Skaja, "How to Interpret Chapter 16 of the *Zhuangzi* 'Repairers of Nature'," in Ames, ed. *Wandering at Ease in the Zhuangzi*, 102–103. Compare also my "Democratic Action and Experience: Dewey's 'Holistic' Pragmatism," in *The Premise of Democracy: Political Agency and Transformation* (Albany, NY: State University of New York Press, 2010), 43–65.

THREE

Thinking at the Edge of Thought

Nothingness, Non-Being, Emptiness

Science wants to know nothing of nothingness. —Martin Heidegger

Apart from other designations, our modern age is also frequently called the "age of science and technology," a designation which appropriately honors the legacy of the great pioneering works of Descartes and Francis Bacon. While the former established the foundations of modern reason or rationality, Bacon provided the basis for the empirical testing of findings and their technical utilization for the benefit of "human comfort." To be sure, the fruits of their works took some time to mature and to penetrate consistently the entire fabric of modern social life. After an intermediary period (termed "Enlightenment") dedicated to the refinement of pure reasoning and cognition, science and technology came finally into their own as the decisive hallmark of the nineteenth century, a hallmark captured by the label "positivism." What the label indicated was that the only acceptable or legitimate type of "knowledge" henceforth was to be knowledge derived from the observation of "positively given" empirical data and the process of rigorous testing and verification. What was left outside the scope of positive knowledge was the entire field of traditional "non-epistemic" insights and teachings which now was pejoratively labeled "meta-physics" or else intuitive speculation. With this demarcation in place, traditional "ontology" fell victim to "critical" rationality, just as traditional theology was sequestered in the narrow domain of private sentiment and belief.

To some extent, the demarcation was not entirely without merit. Like Occam's razor, the positivist critique cut away from thinking a lot of flabby ruminations and fancy illusions. Among the latter was, above all,

41

the assumption of a "higher" world, separate from the empirical world but endowed with the same tangible or "positive" qualities (what is sometimes called "higher realism"). Seen against this background, positivist critique had the distinct advantage of undercutting the traditional "doubling of the world," a doubling which often resulted in the assignment of a robust, triumphalist "positivity" to perennial verities (and thus in a totalized positivism). No doubt, positivist critique also resulted in a widespread sense of loss, often articulated in terms of a growing nihilism or collapse of cultural meaning. However, on a different reading, the loss can also (and more plausibly) be viewed as the negation or nihilation of an illusion, the illusion contained in the "doubling of the world." It was Friedrich Nietzsche—often accused of a destructive nihilism—who alerted us to the salutary effects of the end of "doubling": namely, the transformation of both worlds and of the meaning of "world" as such.[1] What his reminder brings into view is the Achilles heel of "positivism" in every form: its inability to grasp or come to terms with the "non-positive," that is, negativity, nothingness or non-being.

In the following I want to explore the role of negativity or non-being in three contexts: First, in the negative metaphysics or "me-ontology" of Heidegger; next, in the negative or dialectical theology of Paul Tillich; and finally in the broader field of Asian thought, particularly its emphasis on "emptiness."

"DAS NICHTS NICHTET"

In recent Western philosophy, the thinker who wrestled most resolutely with the modern supremacy of science and technology was Martin Heidegger. His primary way of challenging the sway of positivism was to raise again the "question of Being"—a question which had been sidelined by critical (especially Kantian) philosophy as unscientific and illegitimate. By raising the "ontological" question—famously associated with the work of Aristotle—Heidegger exposed himself to the accusation of bringing back into vogue a pre-critical metaphysics of "substances" (and hence a traditional "two-world" theory). How groundless this accusation was could be seen by a glance at his first major publication, titled *Being and Time* (1927), where the Being-question occupied a central place. As Heidegger tried to make abundantly clear, "Being" as used in the text is not a fixed "substance" (in the medieval sense), nor an abstract metaphysical "concept" (as postulated by rationalism). Rather, it designates merely the target or content of an open "question": namely, the question what meaning can be assigned to the "being" predicated of all things and entities in the world. Seen in this light, the Being-question is not raised from "on high" or a "view from nowhere," but precisely from the angle of a particular entity—human being or *Dasein*—for which the sense of its own existence is radically at stake. In Heidegger's words: "From this angle, the Being-question is nothing other than the radicalization of an orientation essentially implicit in human *Dasein*, namely, its pre-ontological understanding of Being [in all its forms]."[2]

Heidegger's comments in *Being and Time* did not entirely silence his critics. Whatever suspicions of a substantive metaphysics remained, however, were completely blown out of the water by a piece he wrote soon after the cited text: the essay "What is Metaphysics?" (1929). As Heidegger initially acknowledges, metaphysics in the traditional sense is indeed something that "comes after" (*meta*) or leads beyond the purely physical or empirical domain. But where does metaphysics lead? Does it lead into a presumed "higher" realm of substances? Following up on his earlier comments, Heidegger stresses the "questioning" aspect of ontology: the aspect which calls into question all beings including the being-status of the questioner or *Dasein* itself. At this point, far from offering a refuge, metaphysics turns into an engine for the internal overturning of positivism in any form: namely, by disclosing the hidden underside of non-positivity or non-being in all positive entities. For Heidegger, this dimension is not accidentally but necessarily shunned by positive science. In his words: From the angle of modern science, "what should be examined are positive beings or entities only, and besides that: nothing; that is, solely empirically given entities, and beyond that: nothing." Pursuing this argument and sharpening the edge of anti-positivism, the essay adds: "If science is right, then only one thing is sure: science wishes to know nothing of the 'nothing' (*das Nichts*)." Differently and still more provocatively put: "Science wants to know nothing of nothingness."[3]

Contrary to the leanings of science, nothingness for Heidegger cannot simply be sidelined or ignored; more importantly, science itself is unable to make a clean break from it. In fact, precisely by focusing entirely on what *is* in opposition to what *is not*, science cannot fail to acknowledge the latter. In Heidegger's words: "Nothingness is sharply rejected by science and given up as a nullity (*das Nichtige*). Yet, when we abandon nothingness in this way, do we not actually acknowledge it?" So, it seems that when science "tries to grasp its proper essence," the positivity of things, "it calls upon the 'nothing' for help"; it "has recourse to what it rejects." Hence, positivism is breached; but what about the nothingness that is being shunned? How can one properly speak about it? Clearly, nothingness is not merely another "something," though perhaps a "something" that is difficult to grasp. In this way, nothingness would simply be leveled into the realm of positivity—and thus be lost. Another misleading way of approaching nothingness is in terms of traditional logic of affirmation and negation. Along this line, nothingness would be captured by the word "not," whereby we say that it is *not* a particular thing—leaving open the possibility that it is another thing. For Heidegger, the universal logic of negation—A is not non-A—does not touch or reach to the level of nothingness. As he observes, when we resort to "not" (*nicht*) or negation (*Verneinung*), nothingness seems to be governed by a higher specification in the sense that it has the status of a "a particular kind of negated entity." Hence, the question arises whether nothingness derives from "not" and "negation"—a purely logical question—or whether it is the other way around. Relying on the Being-question and anticipating further clarification, the essay states firmly: "We assert that nothingness (*das Nichts*) is more original than the 'not' and negation."[4]

If it is not simply a matter of logic, the question remains how nothingness can at all be accessed or experienced—again from the vantage of an entity or being (*Dasein*) for which the meaning of its own being is at stake. Since there is no positivity to cling to, nothingness can only be encountered as a kind of occlusion, withdrawal, or slippage—a slippage not of this or that positive entity but of all entities and of "Being" as such. One way to encounter such slippage is in the mode of radical boredom (*Langeweile*). In Heidegger's words, boredom is still superficial and manageable as long as it concerns "this book or that play" that drags on and is boring. By contrast, radical or profound boredom is something "drifting here and there in the abysses of our existence like a muffling fog, while removing all things and people and oneself into a curious indifference." This kind of boredom discloses "all beings as a whole"—in a state of withdrawal. Another way of encountering nothingness is in a state of dread (*Angst*)— which needs to be distinguished sharply from ordinary fear (*Furcht*). The latter is always related to a particular danger. "We become afraid or fearful," Heidegger writes, "in the face of this or that particular being that threatens us in a particular respect. Hence, fear is in each case of fear of something particular." The situation is entirely different in the case of dread—which actually may be accompanied by calm. "Dread," we read, "is always a dread of . . . but not a dread of this or that particular thing." What happens in dread is that one feels abandoned, uncanny, or "unhomely" (*unheimlich*) because the entire fabric of being slips away or sinks into indifference. This withdrawal of things as a whole produces an oppressive dread because we can get "no hold on things" and this lostness overcomes us in our core. Hence, "dread (again) discloses nothingness."[5]

To be sure, in disclosing itself in dread, nothingness does not show itself forth as a thing, a particular entity, or object: "Dread entails no [cognitive] grasping of nothingness." What happens is rather that beings as a whole become "irrelevant" (*hinfällig*); nothingness shows itself forth "with and in beings as their slipping away as a whole." This slipping away, however, does not by any means denote a destruction or annihilation (*Vernichtung*); rather, it harbors the active and enabling potency of "nihilation" (*Nichtung*). In Heidegger's words: Nothingness "is neither an annihilation of beings nor does it spring from negation (*Verneinung*). Nihilation will not submit to calculation in terms of annihilation and negation." Instead: "nothingness itself 'nothings' or nihilates (*das Nichts nichtet*)" by showing itself forth. By retreating from the whole of beings, nothingness "discloses these beings in their full but hitherto hidden strangeness as the radically other—in relation to nothing." At this point, the crucial capacity of nihilation comes into view: its ability to unleash the Being-question. "In the clear night of the nothingness of dread," we read, "the original openness of beings as such arises: that they are beings and not nothing." In this sense, nihilation functions as the premise or condition of possibility of the Being-question and of the inquiry into beings, an inquiry mandatory for human being or *Dasein*: "The essence of the originally nihilating nothingness lies in this, that it brings *Dasein* for the first time before beings as such. Only on the ground of the original disclosure of nothingness can human existence approach and investigate beings."[6]

The implications of this view of nihilation are enormous: it entails the overturning of a self-sufficient positivism, but also an overturning of a self-contained and other-worldly metaphysics. With regard to positivism, nihilation reveals a hidden underside of knowledge usually denied by scientists: the operation of a non-positive "transcendence" undergirding empirical research (and reflected in the "ek-static" quality of *Dasein*). In terms of the essay: "*Dasein* means: being held out into nothingness." Being held out in this way, "*Dasein* is in each case already beyond beings as a whole. But this being beyond beings we call 'transcendence.' If in its ground *Dasein* were not transcending [or ek-static], it could never be related to beings nor even to itself." This does not mean that human beings are always and in every instant able or willing to face up to the Being-question and to its corollary of nothingness. As Heidegger had already pointed out in *Being and Time*, it is quite possible for *Dasein* to lose itself in shallow everydayness, in time-consuming enterprises and busy affairs. According to the 1929 essay: "We commonly lose ourselves among things and beings in a certain way." And the more we are preoccupied with things and beings, "the less we allow beings as whole to slip away and the more we turn away from nothingness and nihilation." In a time of incessant media bombardment this temptation is nearly overwhelming; once we surrender, "we hasten into the superficial publicity of existence."[7]

The effects of nihilation also touch on the meaning of metaphysics. From Heidegger's vantage, metaphysics does not involve an exodus from the world of beings into another realm peopled with "higher" positivities. Rather, nihilation operates as a "metaphysical" leaven or source of unrest in the world of beings, enabling the constant renewal of the Being-question, including the question of the meaning of *Dasein*. As he states: "For human existence, nothingness makes possible the disclosure of beings as such." Hence, nothingness does not merely constitute the antithesis or "counter-concept" to being; rather "it belongs to the unfolding of

Being as such." More sharply put: "In the Being of beings, the nihilation of nothingness occurs." With this formulation, the traditional notion of metaphysics returns—but now under completely different auspices. As Heidegger acknowledges, traditional metaphysics has always inquired into the status and meaning of beings as a whole. On the basis of a brief historical overview, however, he shows that the tradition has failed to properly raise the Being-question in its connection with nothingness. What the historical review reveals is that nothingness has always tended to be treated simply as "the counter-concept to Being, that is, as its negation." This treatment becomes impossible once nihilation and its potency are taken fully into account. At this point, "nothingness does not remain the indeterminate opposite of beings but reveals itself as belonging to the Being of beings." At the end of his essay, Heidegger returns to the issue of modern science, insisting that, properly construed, science presupposes a nihilating metaphysics—which is not a denial of science: "Only if science exists on the basis of (such a) metaphysics, can science advance in its basic task, which is not to amass and classify bits of data, but to disclose in ever-renewed fashion the entire field of knowledge in nature and history."[8]

The concern with nothingness and nihilation persisted and even intensified in Heidegger's later works—albeit with an important twist. In a sense, the so-called *"Kehre"* (during the 1930s) sought to correct a lingering anthropocentric existentialism in his earlier writings, a feature which was evident in the primacy assigned to human *Dasein* among all possible beings in the world. In the course and as a result of the *Kehre*, Heidegger portrayed *Dasein* not so much as the source and instigator of the Being-question than as the participant and beneficiary of an ongoing disclosure of Being. The change affected other important earlier formulations—or at least shielded them from misinterpretation. Thus, despite its strongly revised character, the term "metaphysics" (used in the 1929 essay) became suspect as presenting merely the outgrowth of a fertile mental imagination. As a corrective, Heidegger came to prefer such terms as "meta-metaphysics" or "end of metaphysics." In the same way, the working of nothingness (*das Nicht nichtet*) came to be presented in novel ways. Thus, instead of being linked closely with such experiences as dread and boredom, nihilation was seen as an essential feature of Being as such, a feature famously captured in such expressions as "oblivion of Being" (*Seinsvergessenheit*) and "abandonment of/by Being" (*Seinsverlassenheit*). This means that, far from being episodic or accidental, non-being is an essential corollary of Being, such that Being reveals itself basically through concealment or shows forth by sheltering itself.

A first indication of the changed outlook can be found in Heidegger's *Contributions to Philosophy: On Ereignis* (1936). In that text, Being is no longer readily accessible to *Dasein* by raising the Being-question; on the contrary, it discloses itself only and essentially in the mode of withdrawal, refusal, or slipping away. What this means is a radical human dispossession and dislocation; in a happening termed *"Ereignis,"* Being grants to all things and beings their proper abode in a basically non-possessive way. This entails that, far from denoting a steadily unfolding teleology, disclosure of Being necessarily implies nihilation and the "going under" (*Untergang*) of the truth of Being and its guardians. In Heidegger's words: "Being itself demands this 'going under'; only by thus going under are beings granted or guided into their own. This guidance or granting is what we call *Ereignis*." A later passage in the same text addressed specifically the issue of nothingness and nihilation. As on previous occasions, Heidegger separates the issue firmly from the operation of a logical negation, asserting that this treatment does not bring us an inch closer to an understanding of Being and nihilation. But, he continues, "how would the situation be, if Being emerges precisely in its withdrawal and discloses itself through refusal? Would such withdrawal amount to a nullity or rather to the most sublime grant? In the latter case, is it not precisely through the nihilation of Being that nothingness (*das Nichts*) acquires that granting potency which lies at the basis of all creation and creativity?" Seen from this perspective, "nihilation must inhabit Being itself and must be the ground for *Ereignis* as non-possession and refusal which is a mode of granting."[9]

Heidegger's reflections on nothingness and nihilation were continued in his treatise *Besinnung*, translated as *Mindfulness*(1938/1939). The text explicitly goes back to the essay "What is Metaphysics?" indicating both its strengths and weaknesses. As Heidegger indicates, the essay tried to go beyond the traditional conception of metaphysics, where the latter involves contemplation of the "beingness" of all beings without proper attention to the presupposed ground of Being and its connection with nothingness. Despite this transgressive intent, however, the essay employed the term metaphysics, but now in the sense of a "genuine metaphysics" or "meta-metaphysics." To gain a more adequate grasp of the issue, the text distinguishes three different ways of conceiving nothingness in its relation to Being. There is, first of all, Hegel's conception of the identity of Being and nothingness, an identity which derives from the empty indeterminacy of both terms. There is, next, the "meta-metaphysical" conception (outlined in 1929) which sees nothingness basically as "nihilation," accessible to *Dasein* in boredom and dread (*Angst*). There is finally the "me-ontological" perspective—inspired by the experience of the *"Kehre"* —which views nothingness as the "un-ground" (*Ab-grund*) sheltering the essence of Being. In the latter perspective, Heidegger comments, nothingness (*das Nichts*) "loses any semblance or superficial impression of a negativity or nullity," because un-ground here means "the refusing withdrawal (*Verweigerung*) seen as the *Ereignung* of a gift."[10]

Besinnung/Mindfulness offers some additional comments which clarify the preceding points. Thus, the move from the second to the third conception just mentioned can also be grasped as a transition from nihilation (*Nichtung*) to refusal (*Verweigerung*). In Heidegger's words: The nihilating nothingness is not self-propelled but derives from "the essence of Being as refusal (or Being's self-sheltering)"; only this refusal "give rise to any kind of negation (*Verneinung*)." Seen from this angle, nothingness emerges as "the un-ground (*Abgrund*) of the clearing granted by refusal. The non-being of the ground derives from the same refusal." In its reformulation after the *Kehre*, the so-called Being-question treats nothingness not as a nullity in comparison with beings, but rather "as *Ereignung*, that is, as a gift granted from the abundance of Being." This gift has long been covered over by traditional metaphysics in its complicity with the modern "age of science and technology." This complicity has for a long time "overpowered" the essence of nothingness, just as it has overpowered "its close connection with Being as the latter's un-ground." The reason for this overpowering can be found in the "abandonment of beings by Being" (*Seinsverlassenheit*) in the modern age as a result of the primacy granted to fabrication, technical making, and production (*Machenschaft*).[11]

TILLICH ON DIVINE NON-BEING

Heidegger's thoughts on nothingness and nihilation have had widespread repercussions in many intellectual circles. Somewhat surprisingly, they also reached the often cloistered world of Christian theologians, especially those who had already been stirred from intellectual slumber by Nietzsche's word about the "death of God." A prominent figure among the latter group was the Protestant theologian Paul Tillich who became acquainted with Heidegger during the years they both shared at the University of Marburg. Born in 1886 as the son of a Lutheran pastor, Tillich grew up in northern Germany during the Bismarck era. He studied philosophy and theology at the universities of Berlin, Tübingen , and Wittenberg, graduated with two degrees from Breslau and Wittenberg, and was ordained as a Lutheran minister in 1912. During World War I, he served as a chaplain in the Imperial Army. After the war, his academic career began in Berlin from where he moved to Marburg (1924–1925) where he started to develop his systematic theology but also absorbed a good deal of the "existentialist" spirit of the Weimar years. Subsequently, Tillich served as theology professor in Dresden, Leipzig, and finally at the University of Frankfurt. Attracted to the "Confessing" Lutheran Church, he went on collision course with the Nazi movement and, in 1933, was dismissed from his position and forced to emigrate. With the help of Reinhold Niebuhr he joined the faculty at Union Theological Seminary in New York where he taught until 1955, publishing some of his major works and gaining the reputation as a leading Christian theologian. Toward the end of his life, he had spent a decade at Harvard before he passed away in 1965.

From the time of his arrival in New York throughout the Second World War, Tillich was active both academically and politically, beaming a long series of anti-fascist broadcasts from America to Germany. The real flourishing of his academic as well as his pastoral work, however, came after the war. In 1948 he published several volumes of his pastoral sermons, books which gave him an audience far beyond academia. His most impressive scholarly achievement came in 1951 with the publication of the first volume of *Systematic Theology*. The text is an impressively innovative or pioneering work, by combining in a rigorous fashion the tenets of traditional Christian theology with the insights of existentialism and especially some of the teachings of Heidegger. Basically the combination yielded a framework which does not impose Christian dogmas "top down" on human believers, but rather starts "bottom up" from concrete human experiences and agonies—while not discarding the guiding capacity of revelation. Thus, to be alive and relevant, Christian theology for Tillich has to start from the human "condition" or "situation" in all its fallenness and finitude; at the same time, this condition is not closed in upon itself, but serves as a springboard for the ever-renewed human quest for the meaning of existence. In his words: Substantial theology "must consider the creative interpretation of existence, an interpretation which is carried on in every period of history." Seen from this angle, situation and revelation (or *kerygma*) are locked in a mode of "correlation": a correlation of question and response (familiar from the hermeneutics of Heidegger and Hans-Georg Gadamer). Inserted into this mode, theology "answers the questions implied in the 'situation' in the light of the kerygmatic message and with the means provided by the situation whose questions it answers."[12]

As one should note, attentiveness to the human condition for Tillich does not mean a surrender to social conditioning or a myopic anthropocentrism. Just as for Heidegger, human being (*Dasein*) is basically openended, ekstatic and, as such, standing out into the Being-question or the mystery of Being. Following this line of approach, *Systematic Theology* traces the Being-question basically to human existence (rather than Being as such): "Revelation answers questions which have been asked and always will be asked because they are 'we ourselves.' Man [*Dasein*] is the question he asks about himself, before any question has yet been formulated." Using directly Heideggerian language, Tillich continues: "Being human means asking the question of one's own being and living under the impact of the answers given to this question." This outlook sheds additional light on the method of "correlation" mentioned before—and actually transforms it into a kind of symbiosis. Viewed under these auspices, we read, systematic theology "makes an analysis of the human situation out of which the existential questions arise, and it demonstrates that the symbols used in the Christian message are the answers [or possible answers] to these questions. The analysis of the human situation is done here in terms which today are called 'existential'" —although the issue is "much older than existentialism." Tillich in this context refers to "the mystical identification of the ground of Being with the ground of self," the "use of psychic categories for ontological purposes in Paracelsus, Böhme, Schelling," and "Heidegger's notion of '*Dasein*' as the mode of human existence and the gateway to ontology."[13]

Given the ecstatic openness or outreach of human *Dasein*, the question arises regarding the target of this outreach or quest. For the theologian, the ultimate target is divine revelation or message (*kerygma*), which leads to the further question concerning the nature or source of the revelatory disclosure. Together with Heidegger, Tillich rejects the terms "substance" and "cause" as designations of this source, because both imply a predictable process and outcome. Again together with the German philosopher, he prefers to speak of the ultimate "ground of Being"—identifying the latter with "God." As he states: "The religious word for what is called the ground of Being is God. . . . A doctrine of God as the ground of revelation presupposes the doctrine of Being and God, which on the other hand is dependent on the doctrine of revelation." This manner of speaking, however, conjures up the basic "ontological" question: "What is Being itself? What is that which is not a particular being or group of beings . . . but rather something which is always thought implicitly, and sometimes explicitly, if something is said to *be*?" In pursuing this question, the inquirer faces an abyss or un-ground (*Abgrund*). "The ontological question," Tillich writes, "arises in something like a 'metaphysical shock'—the shock of possible non-being." What renders this shock so powerful or dramatic is the fact that "everything disappears in the abyss of possible non-being; even a god should disappear if he were not Being itself."[14]

In *Systematic Theology*, the shock of non-being is directly linked with the ecstatic quality of *Dasein*. In Tillich's words: "Only man can ask the ontological question because he alone is able to look beyond the limits of his own being and of every other being." Differently put: "Man is able to take this standpoint because he is free to transcend every given reality. He is not bound by 'beingness' [*Seiendes*], but can envisage nothingness and ask the ontological question." The question remains regarding the meaning or status of nothingness or non-being. Again in line with Heideggerian insights, the issue of non-being cannot be reduced to an aspect of universal logic (A cannot be non-A). "One can ask," Tillich states, "whether non-being is anything more than the content of a logical judgment, a judgment in which a possible or real assertion is denied." What is wrong about this reduction to negation or logical negativity is that the non-being inheres in the "ontological structure" of *Dasein* which stands out beyond beings. Due to this ontological structure, *Dasein* must be able to be "separated from its [ontic] being in a way which enables it to look at it as something strange and questionable." By virtue of this separation, man (*Dasein*) "participates not only in Being but also in non-being." Differently expressed: "The very structure which makes negative judgments possible proves the ontological character of non-being. Unless *Dasein* participates in non-being, no negative judgments are possible; in fact, no judgments of any kind are possible. This mystery of non-being cannot be solved by transforming it into a kind of logical judgement."[15]

As should be noted, non-being for Tillich is not simply the antithesis of Being, located in some realm outside ontology. He speaks of a "dialectical participation of non-being in Being" and also points to the "recent rediscovery of the Being-question" in the revival of pre-Socratic philosophy and the work of Heidegger. In a broader historical context, he adds, the issue of non-being had been raised by negative theology, the theology of the *via negativa* where non-being meant "not being anything in particular" and thus referred to "being everything" and ultimately to Being itself. In Christian mysticism, a similar issue had been addressed and formulated in terms of a "dialectical" participation. The text here refers to Jakob Böhme's and Meister Eckhart's "*Ungrund*," Schelling's first potency, and Berdyaev's "me-ontic freedom." The most forceful formulation, however, was provided by recent existential philosophy with the idea of an "encountered nothingness." As exemplary, Heidegger's work is lifted up again, especially his notion of "nihilating nothingness" (*das Nichts nichtet*) which portrays *Dasein* as being "threatened by non-being in an ultimately inescapable way, that is, by death." This anticipation of non-being in death gives to human existence its dramatically "existential character."[16]

The dialectical participation of non-being in Being was Tillich's signature formulation, and he remained faithful to it in his subsequent publications. Easily his most influential and popular writing was titled *The Courage to Be*, published just a year after *Systematic Theology*. Written in accessible style for a general audience, the book reveals more clearly than the earlier work the decisive "existentialist" contours of Tillich's outlook. Basically, the "courage to be" was meant as an antidote to the nihilating nothingness threatening the human condition or situation. "Courage," we read, "is self-affirmation 'in-spite-of,' that is, in spite of that which tends to prevent the self from affirming itself." To be sure, the threat of nihilation cannot be shunned or ignored, but it can be met and faced down by the realization that non-being is a corollary of Being. "If one asks," the book continues, "how non-being is related to Being itself, one can only answer metaphorically: Being 'embraces' both itself and non-being, or Being has non-being 'within' itself"—that is, non-being participates (dialectically) in Being. Basically, in Tillich's view, nothingness or non-being threatens *Dasein* by engendering anxiety (*Angst*) about fate and death; such anxiety, in turn, endangers human "spiritual self-affirmation" by inflicting on *Dasein* the experiences of "emptiness and meaninglessness"—where the latter poses "the absolute threat of non-being" while emptiness negates "the special contents of spiritual life." The situation is not totally hopeless or inescapable, however, since non-being remains "dependent on the Being it negates"—which points to the "ontological priority of Being over non-being." For Tillich, "there could be no negation if there were no preceding affirmation to be negated." The central human predicament, hence, is how to re-affirm the prior affirmation—an effort which acquires the "courage to be."[17]

In lieu of pursuing the theme further in Tillich's later writings, it seems opportune to lift up briefly some of the distinctive accents of his treatment, especially as they compare with Heidegger's work. As indicated, these accents are basically "existentialist" in character, with a focus on self-affirmation. When invoking the philosopher, the references are exclusively to the early Heidegger, never to the later thinking after the "*Kehre*." As we read in *The Courage to Be*, Heidegger in *Being and Time*—which has "its independent philosophical standing whatever Heidegger may say about it [later] in criticism and retraction"—describes "the courage of despair in philosophically exact terms," elaborating "the concepts of non-being, finitude, anxiety, care, having to die, guilt" and so on. The same text portrays Heidegger's "historical function" as the effort "to carry through the existentialist analysis of the courage to be as oneself more fully than anyone else." This statement is followed by a comparison with Jean-Paul Sartre who, we are told, drew "the consequences from the earlier Heidegger"—although consequences "which the later Heidegger did not accept" (and could not accept given the later development of his thought). Other points of demarcation might also be noted. Thus despite his refusal (with Heidegger) to reduce non-being to a logical operation, many of Tillich's comments tend precisely to support such a reduction—as when he presents non-being as "dependent on the Being it negates" or when Being is equated with "the [Hegelian] negation of the negation." The same tendency is evident in his notion of "dialectical participation," seeing that dialectics means the overcoming of mutually opposing positions. The entire argument of *The Courage to Be* draws its strength and plausibility from the opposition between negation and affirmation, between self and other or self and the world—which is quite distant from Heidegger's conception of *Dasein* as "being-in-the-world."[18]

NON-BEING AND EMPTINESS

Curiously, as Heidegger in the "*Kehre*" was steadily turning away from existentialist "self-affirmation," he was moving closer and closer to Asian thought (the latter seen as philosophy and not a simple "mysticism" devoid of thought). As we know, after the war Heidegger was collaborating with a Chinese philosopher on a German translation of the *Daodej-*

ing—an effort which did not reach fruition. Independently of this particular effort, his reformulation of the Being-question in terms of withdrawal, slippage or sheltering shows an unmistakable affinity with the South and East Asian notions of no-self (*anatman*) and emptiness (*sunyata*)—notions which are far removed from the logic of negation (and the negation of the negation). These notions are central in the traditions of Buddhism (especially in its Mahayana version) and Chinese Daoism—traditions which, though originally quite distinct, came to interact with and influence each other in the course of centuries. What links the two traditions is above all their move "beyond metaphysics" in the direction of something like a "meta-metaphysics" or a "me-ontic" ontology. Another way to characterize this move is in terms of a practical-ethical disposition—in the sense that the "overcoming" of metaphysics is crucially linked with the effort of self-overcoming or "self-emptying" in favor of a non-possessive and mindful praxis.

The emphasis on self-emptying was always pronounced in Buddhist thought. As we know, Buddhism in many ways arose as a reaction to traditional Hinduism which tended to assign a fixed substance or "essence" to all beings, including the human self. In opposition to this kind of "essentialism," Buddhism insisted from the beginning on the emptiness of selfhood (*anatman*) and on the lack of a fixed substance in all beings. One result of this opposition was the collapse of the Hindu caste system, now seen as an arbitrary social construct. The danger of this denial of essences was the lapse into radical randomness or nihilism—an alternative functioning as a "counter-essentialism" (and thus still obeying the logic of negation). Aware of this logical trap, leading Buddhist thinkers (like Nagarjuna) always rejected the binary opposition between essentialism and nihilation, thus charting a way toward a practical "meta-metaphysics" (for instance, by speaking of a "middle-way" which is neither a compromise nor a synthesis). What needs to be remembered at this point is the "liberating" quality of Buddhism, its attempt to overcome both essentialist fixation *and* nihilist despair along a path which leads to the self-emptying of emptiness and to a freedom beyond being and its lack. It is this freedom which is celebrated in these famous lines of the *Dhammapada* (ca. 300 BCE): "Ah so pleasantly we live/without enmity among those with enmity. Among humans who hate/we dwell with friendly compassion."[19]

The verses in the *Dhammapada* do not depart from, but celebrate and underscore central Buddhist teachings. As is well known, the Buddha Shakyamuni himself was extremely weary of metaphysics, preferring to accentuate the value of rightful living or (what I call) "mindful praxis." This accent is evident in his sermons, many of which have been collected in the so-called *"Prajna-paramita"* Sutras (ca. 200 BCE). Here are some examples. When asked by a follower what "enlightenment" or "liberation" means, the Buddha replied: "Enlightenment happens when all things are seen in their intrinsic empty nature, their 'suchness,' their un-graspable wonder. Names are incidental, but that state which sees no division, no [subject-object] duality, is enlightenment." The same empha-sis on non-duality, on the non-oppositional entwinement of all things is also present in another sermon which states forcefully: "You cannot ex-pect to have the conditioned (*samsara*) without the unconditioned (*nirva-na*); nor can you expect to have the unconditioned without the condi-tioned—*nirvana* apart from the world. The wise person *lives* this truth and, by the same token, engages in good works." And what does such a meta-metaphysical engagement mean? "Without ever swerving from the realization that all things are empty of fixed self-nature (*svabhava*), the wise person inspires or assists all beings to free themselves from the yoke of self-nature." From the angle of this freedom, the world emerges as a "new and wondrous creation." In the same vein, another passage from the Sutras deserves to be lifted up: "Empty, calm and devoid of self-being [self-centeredness] is the nature of all things; no particular being exists enclosed by itself . . . All beings in the world, beyond the limitations of words, dwell in their pure nature in the infinity of space."[20]

Although learnings from different cultural and historical back-grounds, classical Daoist texts basically concur with Buddhist teachings on the issue of emptiness, non-duality, and non-self-centeredness. Re-garding non-duality and non-opposition, the *Daodejing* states in an open-ing chapter: "Being and non-being create each other; difficult and easy support each other, high and low depend on each other." The practical orientation of this "meta-metaphysics" emerges in the next chapter which states: "The sage person leads by emptying people's minds but filling their cores, by weakening their selfish ambition but toughening their resolve." The point of non-metaphysical conduct is to follow the *Dao* (Way) which itself is beyond epistemic duality. In terms of one of the later chapters: "*Dao* is indistinct and ineffable, yet it contains images or forms; though indistinct and ineffable, yet it embraces events . . . Since before time and space were, the *Dao* 'is.' It is beyond *is* and *is not.*" As one should note, "is" and "is not," being and non-being, here are not related in the mode of logical negation, but rather interpenetrate each other, thus allowing for releasement or liberation from restrictive cognitive catego-ries. This point is underscored in another chapter of the *Daodejing* where we read: "To be aware of the positive, yet to abide in the negative is to be the abyss of the [cognitive] universe . . . To be the chasm of the universe means to have attained realization and to abide in the state of original non-differentiation [or non-self-nature]."[21]

The practical, post-metaphysical orientation is fully maintained in the text called *Zhuangzi* (4th century BCE) with its emphasis on the "no-self" or emptiness of selfhood. As we read in a famous passage: "The genuine person has no self (*wu ji*), the spiritual person has no merit (*wu gong*), and the sage has no name (*wu ming*)." To the extent, that selfhood is associated with an inner or self-possessed mind, the *Zhuangzi* wants to have little to do with it, speaking of the desirability of "letting the mind (*xin*) wander" and even of "mind-fasting" (*xinhai*). In terms of chapter 4 of the text: "Listen not with your ears, but with your heart-mind (*xin*); listen not with your heart-mind but with your primal breath (*qi*) . . . That breath awaits things emptily. It is only by wandering along the way (*Dao*) that one can gather emptiness, and emptiness is mind-fasting." And chapter 11 portrays releasement as a state when "the eyes see 'nothing,' the ears hear 'nothing,' and the mind knows 'nothing' (*xin wusuozhi*)."

In these lines, the withdrawal or slippage of ultimate reality (or Being) announces itself, together with the oblivion or forgetting of epistemic "knowledge." However, none of these comments counsel carelessness or retreat into self-interest. According to chapter 11, the teaching of the sage is "like the shadow of a form, like the echo of a sound. When questioned, he responds, sharing his thoughts freely and serving as the companion of all under heaven . . . He leads the teeming masses back to where they belong, to wander in limitlessness. He passes in and out of non-attachment." The gist of this unobtrusive, non-self-centered mind and non-manipulative conduct is also reflected in chapter 6 which explicitly refers to Confucius: "'Fish delight in water,' said Master Kong, 'and man delights in the Way' . . . Therefore, it is said, 'fish forget themselves in the river and lakes; humans forget themselves in the practice of the Way.'"[22]

NOTES

1. Epigraph is taken from Martin Heidegger, *Was ist Metaphysik?* (1929). My translation. See Friedrich Nietzsche, "How the 'True World' Finally Became a Fable," in *Twilight of the Idols*, in Water Kaufmann, ed., *The Portable Nietzsche* (New York: Viking Press, 1968), 485–486, especially the statement (486): "With the true world we have also abolished the apparent one."

2. Martin Heidegger, *Sein und Zeit*, 11th ed. (Tübingen: Niemeyer, 1967), 15; *Being and Time*, trans. John Macquarrie and Edward Robinson (New York: Harper & Row, 1962), 35.

3. Heidegger, "Was ist Metaphysik?" in *Wegmarken* (Frankfurt-Main: Klostermann, 1967), 3–4; "What is Metaphysics?" in David F. Krell, ed., *Martin Heidegger: Basic Writings* (New York: Harper & Row, 1977), 97–98. The essay was presented as inaugural lecture at the University of Freiburg, a lecture attended by many prominent scientists.

4. "Was ist Metaphysik?" 3–6; "What is Metaphysics?" 97–99.

5. "Was ist Metaphysik?" 7–9; "What is Metaphysics?" 101–103.

6. "Was ist Metaphysik?" 10–12; "What is Metaphysics?" 104–105.

7. "Was ist Metaphysik?" 12–13; "What is Metaphysics?" 105–106.

8. "Was ist Metaphysik?" 12, 15–18; "What is Metaphysics?" 106, 109–111.

9. Heidegger, *Beiträge zur Philosophie (Vom Ereignis)*, ed. Friedrich-Wilhelm von Herrmann (*Gesamtausgabe*, vol. 65; Frankfurt-Main, 1989), 7, 246–247. For some English translations (not followed here) see *Contributions to Philosophy (From Enowning)*, trans. Parvis Emad and Kenneth Maly (Bloomington, IN: Indiana University Press, 1999), 6, 173–174; *Contributions to Philosophy (Of the Event)*, trans. Richard Rojcewicz and Daniela Vallega-Neu (Bloomington, IN: Indiana University Press, 2012), 9, 193–194. The

translations are extremely cumbersome and barely readable. "Enowning" is a mere transliteration, and hence obfuscation, of *Ereignis*. The term "Event" entirely lacks specificity of content.

10. Heidegger, *Besinnung*, ed. Friedrich-Wilhelm von Herrmann (*Gesamtausgabe*, vol. 66; Frankfurt-Main: Klostermann, 1997), 312, 375–376; *Mindfulness*, trans. Parvis Emad and Thomas Kalary (New York: Continuum, 2006), 277–278, 333–334.

11. *Besinnung*, 312–313, 376–377; *Mindfulness*, 277–279, 334–335. With regard to "*Machenschaft*" compare my "Resisting Totalizing Uniformity: Martin Heidegger on *Macht* and "*Machenschaft*" in *Achieving Our World: Toward a Global and Plural Democracy* (Lanham, MD: Rowman & Littlefield, 2001), 189–209.

12. Paul Tillich, *Systematic Theology*, vol. 1: *Reason and Revelation, Being and God* (Chicago: University of Chicago Press, 1951), 4, 6. As he states (4): "The 'situation' to which theology must respond is the totality of man's creative self-interpretation in a special period." In pursuing this path, theology for Tillich avoids the derailments of both relativism and fundamentalist absolutism (5): Without the anchor of *kerygma*, "theology would lose itself in the relativities of the 'situation'; it would become a situation itself—for instance, the religious nationalism of the so-called 'German Christians' and the religious progressivism of secular humanists in America . . . [On the other hand,] the pole called 'situation' cannot be neglected in theology without dangerous consequences. Only a courageous participation in the 'situation,' that is, in all the various cultural forms which express man's interpretation of his existence, can overcome the present oscillation of kerygmatic theology between the freedom implied in the genuine kerygma and its orthodox fixation."

13. *Systematic Theology*, vol. 1, 62, with note 19. For some reason, he does not refer to the Eastern Orthodox notion of "*theosis*."

14. *Systematic Theology*, vol. 1, 156, 163–164.

15. *Systematic Theology*, vol. 1, 186–187. In the citations above I have repeatedly substituted "*Dasein*" for "man" in order to avoid the awkward gender bias.

16. *Systematic Theology*, vol. 1, 187–189.

17. Tillich, *The Courage to Be*, 2nd ed. (New Haven, CT: Yale University Press, 1952/ 1980), 32, 34, 40, 46–47.

18. *The Courage to Be*, 148–149. Existentialism or the "existentialist point of view" is presented in the text as the opposite or negation of essence or essentialism (125–128). Regarding non-being and negation compare this comment (179): "Non-being belongs to Being; it cannot be separated from it. We could not even think 'Being' without a double negation: Being must be thought as the negation of the negation of Being" (clearly a Hegelian move, although Hegel is classified as an "essentialist" thinker, 135). A further demarcation resides in Tillich's equation of Being with God—albeit a very non-essential God: a "God above God," or a "God above the God of theism" (186–187).

19. See *The Dhammapada*, trans. John Ross Carter and Mahinda Palihawadana (New York: Oxford University Press, 1987), chapter 15, 254. The notion of the "middle way" (*madhyamika*) is usually associated with the great Buddhist thinker Nagarjuna (2nd or 3rd century CE). In the words of the Japanese Buddhist philosopher Masao Abe: "Nagarjuna rejected as illusory not only the 'essentialist' view which took phenomena to be real [substantial] just as they are, but also the opposite 'nihilistic' view that emptiness and non-being are true reality . . . For Nagarjuna, emptiness was not 'nonbeing' but rather 'wondrous Being.' Precisely because it is emptiness which 'empties' even emptiness, true emptiness is absolute reality which makes all phenomena, all beings, truly *be*." See Abe, *Zen and Western Thought*, ed. William R. LaFleur (Honolulu, HI: University of Hawaii Press, 1985), 94. The book contains an instructive chapter on "Tillich from a Buddhist Point of View," 171–185. References to Heidegger are recurrent throughout the text.

20. See *The Buddha Speaks*, ed. Anne Bancroft (Boston: Shambala, 2010), 28, 33, 121. Compare also *Die Lehre des Buddha – The Teaching of Buddha*, 26th ed. (Tokyo: Kosaida Printing, 2006).

21. The above citations are taken from *Daodejing*, Chapter 2, 3, 21, and 28. I have consulted and merged together several translations: Stephen Mitchell, *Tao Te Ching: A New English Version* (New York, Harper Perennial, 1988); *Lao Tzu: Tao Te Ching*, trans. D. C. Lau (New York: Penguin Books, 1963), *The Way of Life According to Lao Tzu*, trans. Witter Bynner (New York: Perigee Books, 1972); and *Daodejing "Making This Life Significant": A Philosophical Translation*, by Roger T. Ames and David L. Hall (New York: Ballantine Books, 2003). Commenting on Chapter 11, Ames and Hall observe (91): "The classical Western notion of 'Being' used in a metaphysical sense is generally associated with ontological ground. . . . .while 'non-being' is its strict negation. The Chinese existential verb *you*, by contrast, overlaps with the sense of 'having' rather than the copular, and therefore means 'to be present' or 'to be around.' *Wu* translated as 'nothing' means 'not to be present' or 'not to be around' . . . *You* and *wu* are not ontological categories with all the philosophical implications that would entail, but are rather interdependent categories of 'something' and 'nothing,' of presence and absence." I would prefer to rephrase the last comment in terms of practical conduct: as "presenting" and "absenting."

22. *Zhuangzi*, Chapter 1, 4, 6, 11. I have used these texts: Burton Watson, *The Complete Works of Chuang Tzu* (New York: Columbia University Press, 1968); A.C. Graham, *Chuang-Tzu: The Inner Chapters* (London: Unwin, 1989); and Roger T. Ames, ed., *Wandering at Ease in the Zhuangzi* (Albany, NY: State University of New York Press, 1998). Compare also Jee Loo Liu and Douglas L. Berger, eds., *Nothingness in Asian Philosophy* (New York: Routledge, 2014).

FOUR

Mindfulness and Language

But Only Speak the Word

On the stage as in reality, monologue precedes death. —Albert Camus

Hardly a day passes without news of disasters or calamities. Some of these disasters are natural (like earthquakes or tsunamis); but many or most are man-made or humanly induced. What is one to think of genocides, ethnic cleansings, sectarian bloodbaths, and also of long-distance, push-button killings of anonymous people suspected (without even a shadow of proof) of being at odds with the dominant constellations of power? Can one still say that these ills are "humanly" induced—given that the perpetrators often lack any traces of a human face? So, what has transformed presumably human beings into bloodthirsty monsters, into hideous gigantic machines spewing forth at every step devastating fires, destruction, and death? Who or what has cast this spell on agents, distorting their human features into the bloated grotesque image of bloodthirsty beasts equipped with deadly horns and rapacious fangs? How can the spell be broken or reversed? In fact, can it at all be broken—or is it already too late? And what kind of incantation or exorcism would be needed to overcome the spell? And who in our time is capable or qualified to speak the healing words?

The spell referred to here is not just a showpiece—perhaps an entertaining showpiece—in a theater of the absurd. Rather it is a malady or calamity afflicting all of humanity, perpetrators as well as their victims. It is a festering wound crying out for relief. Nietzsche has captured this wound in Zarathustra's "Midnight Song": "O man, take care (*O Mensch, gib acht*). What does deep midnight declare? (*Was spricht die tiefe Mitternacht?*)" Paraphrasing we might ask: what does deep midnight wish to

59

convey to us, what should we be attentive to or careful about, as we awaken from a deep dream? It is this: "The world is deep (*die Welt ist tief*), deeper than the day had thought (*tiefer als der Tag gedacht*). Deep is its wound or woe (*tief ist ihr Weh*)." And also this: "Woe says: cease, go away (*Weh spricht: vergeh*)." But here the question arises: How can the wound cease or go away? How can it change from suffering to joy or happiness (*Lust*)?[1] How can the spell of agony be broken and make room for redemption? The "Midnight Song" does not directly disclose that secret. Maybe it is best to stop here and ponder the issue for a while. Perhaps we shall find hints in some other statements or utterances or teachings. But we need to move slowly in that direction.

TALES FROM ARABIAN NIGHTS

There is a famous book containing a large collection of night songs or night-time narratives. The collection dates back some twelve hundred years to the time of the Sassanid dynasty in Persia, the time of Harun al-Rashid and the Abbasid caliphate in Baghdad; it is called *Arabian Nights* (and sometimes also *The Thousand and One Nights*).[2] Actually, many of the stories collected in the volume are still older and trace their origins to Indian, Southeast Asian, Mesopotamian, and Arabian sources. To this extent, the book is a treasure chest of premodern, intercultural literary imagination and cross-fertilization. Although coming to us from premodern times, the stories of the book are by no means "untimely" or out of date; they still provide us with ample food for thought and rich nourishment for the heart and the travails of everyday living. Many of the stories deal with unusual happenings (unusual for modern sensibilities): we encounter a panoply of spells—good spells and evil spells—and diverse means of breaking or removing the spells. We also find multiple cross-overs between dimensions of life which in modernity are neatly segregated: the domains of humans, animals, spirits, and divinities (or the divine).

One such cross-over story which occurs early in the volume is "The Tale of the Ox and the Donkey." The story deals with a wealthy merchant who employed many servants and also owned many camels and other animals. The merchant was under a kind of mixed spell, partly good and partly bad: the good part was that he was taught and could understand the language of beasts; the bad part was that, if he revealed this secret knowledge to anyone, he would die. At one time, the merchant overheard a conversation between two of his animals, an ox and a donkey, with the ox complaining that he was forced to work hard all day while the donkey could stay home and enjoy food in comfort. Moved by compassion, the donkey advised the ox to pretend to be ill and unable to work the next day—an advice the ox gladly followed. The merchant, however—being aware of their conversation—simply reversed the animals' roles and placed the yoke of the plow on the donkey, to the latter's great regret. In order to prevent the ox from continuing the charade of illness, the donkey resorted to a ruse: claiming that the merchant had loudly stated that, in case of continuing illness, the ox might as well be slaughtered. On overhearing this clever ruse, the merchant burst out into uncontrolled laughter—prompting his wife to inquire about the reason of this outburst. Yet, realizing that disclosing his secret would result in his death, the merchant refused to answer—prompting his wife to be only more persistent in her demand. At this point, the merchant was rescued by another animal conversation, this time between a dog and a rooster— with the rooster advising the dog on how to treat a wife with firmness and dexterity. Thus, while knowing the language of animals almost led the merchant into disaster, he was saved in the end by the same language skill and the ability of understanding.[3]

Arabian Nights contains many such cross-over stories where "cross-over" often carries the connotation of cross-cultural and inter-faith relations. What one realizes in reading the stories is that the so-called "Orient" at that time was a vast tapestry of interlocking orientations not yet neatly parceled out into competing ethnicities or nationalities. One example is the tale of "The Fisherman and the Demon" where—having been rescued from a bad fate—the demon leads the fisherman over a mountain to a beautiful lake and instructs him to fish there. In pulling out his net, the fisherman discovers fish of four colors: white, red, blue, and yellow. Expecting a reward, he brings the fish to the king of that region who orders them to be fried and served. However, during the cooking preparations, the kitchen wall is split open and discloses a beautiful young woman who says "O fish, have you kept your promise?" To which the fish respond: "Yes! If you return, we shall return; if you keep your vow, we shall keep ours; and if you forsake us, we shall be even." Confused by this experience, the cook spoils the food—which prompts the king to ask for a new batch of fish whose preparation leads to the same uncanny result. After the third failed attempt, the king decides to investigate and, traveling in the direction of the lake, be comes upon a deserted palace inhabited only by a paraplegic prince. As it happens, the prince by mistake had earlier married a wicked she-demon who—out of some spite— had cast a spell on him making him ill and also cast a spell on his city transforming it into a lake. That city—the king learns—had been inhabited earlier by four religious sects: Muslims, Zoroastrians, Christians, and Jews; but the demon's spell had changed the sects into fish: the Muslims white, the Zoroastrians red, the Christians blue, and the Jews yellow. By returning to his engulfed city and by keeping his "vow" to its inhabitants, the prince is finally able to break the demon's spell and to turn the fish back into humans.[4]

One of the most amazing and appealing inter-faith narratives is "The Story of the Jolly Hunchback." In this story, a tailor and his wife one day go out for a stroll in a park and there encounter a hunchback who is joyfully singing, dancing, and amusing himself. Pleased by this encounter, the couple invites the hunchback for supper and places before him a large amount of food. As it happens, overly eager and inebriated, the guest eats too quickly—with the result that a bone gets stuck in his throat and chokes him. In an effort to dislodge the bone, the tailor boxes his chest—which leads to the guest's sudden demise. Afraid of the consequences of their action, the couple decides to get rid of the body, and they do so by placing it at night on the stairway of the nearby house of a Jewish physician. When the physician descends that stairway, he stumbles in the dark over the hunchback's body—who rolls down and lies lifeless at the bottom of the stairs. Afraid of the consequences, the doctor lowers the body through the roof into the house of a Muslim neighbor, propping him up against the wall. Returning home and discovering a dark figure in his house, the Muslim takes a club and strikes the hunchback—who falls down into a lifeless heap. Fearing again the consequences, the Muslim carries the body at night to the market and places it at the shop of a Christian tradesman. Returning to his shop, the merchant mistakes the hunchback for a thief and proceeds to kick and choke him. At this point, however, a watchman (or policeman) arrives at the scene and, finding the hunchback dead, takes the merchant as the murderer to the king—who orders him to be hanged. What is noteworthy here is not so much the strange sequence of events, but what follows. For, when the Christian merchant is just about to be executed, the Muslim steps out of the crowd proclaiming that he was the real murderer deserving to be hanged, whereupon the Jewish doctor steps forward claiming the same thing, whereupon the tailor in turn steps forward. In the end, knowing the hunchback's bad eating habits, the king releases all of them—which is a happy outcome, made even more memorable when compared with our present times (when every one of the accused would quickly put the blame on the others).[5]

The most noteworthy and uplifting lesson found in *Arabian Nights*, of course, is not so much any individual story but the narrative which frames all of the stories: that of king Shahrayar and his vizier's daughter named Shahrazad (or Sheharazad). The king had been horribly betrayed by his wife and a number of other women. Thus, he swore to himself to marry in the future for one night only and to kill the woman the very next morning. This practice king Shahrayar carried on for some time and, in the process, decimated the number of nubile women in his kingdom. At this point, the vizier's daughter steps forward, in the hope of putting an end to the practice and ultimately save the city. Against the strong objections of her father, she offers herself in marriage to the king—who, seeing her beauty, gladly accepts. One condition, however, that Shahrazad made was the presence of her sister Dinarzad in the palace during the night. So, after the king and his new wife had been united, Dinarzad— prompted by the latter—steps forward saying: "Sister, if you are not sleepy, tell us one of your little tales to while away the night until daybreak." In response to this invitation, Shahrazad recounts an amazingly intriguing story, but breaks off just before the climax or denouement— which prompts the king to request a continuation the following night. Shahrazad's skillful narrations keep the king entranced for a thousand nights—in the course of which he comes to trust and love her, thus in the end sparing her life and keeping her as his queen. The overall lesson is surely memorable for all times: namely, that speech and narration have a humanizing effect and, in the end, save humanity and the city. In this respect, *Arabian Nights* corroborates Hölderlin's famous lines: "Much has humanity learned, invoking divine names, since we are conversing and can listen to one another."[6]

OUR PLEASANT SOJOURN IN ARANJUEZ

The humanizing and elevating role of language and narration is not limited to a distant "Oriental" past. The lesson is valid also in modernity and in our contemporary period—when it is too frequently neglected or shunned. Not long ago, I had the pleasure of attending in Vienna the performance of a play by the poet and playwright Peter Handke titled *Die schönen Tage von Aranjuez* (Our Pleasant Sojourn in Aranjuez). The title of the play is unusual and intrigued me. In effect, it was both the (well deserved) reputation of Handke as a writer and the title of his play which brought me to the theater. The title comes from the opening line of Friedrich Schiller's drama *Don Carlos*. There, the priest Domingo says to Don Carlos, the son of King Philip II of Spain: "Our pleasant sojourn in Aranjuez is over now; your Highness leaves the place no happier than before" (*Die schönen Tage in Aranjuez sind nun zu Ende; eure königliche Hoheit verlassen es nicht heiterer*).[7] Schiller's drama is a great, classical play in the tradition of Greek tragedy. Thematically, it is a paean to human nobility—at least to a *possible* human nobility achieved through the cultivation of virtue and the free acceptance of civic responsibility. Matching its guiding theme, the drama's language is also of great purity and nobility seeking to elevate its readers or audience through the simple beauty of its measured lines. To my relief and delight, I noted that, in adopting Schiller's opening line, Handke did not betray Schiller's language and kept his play throughout free from vulgarity and some corrupted popular tastes.

On the latter score, my delight was in no small measure influenced by the malaise of the contemporary German theater. The malaise is hard to describe and even harder to make credible unless one has experienced it first-hand. The fact is that, on the German stage today, some of greatest German plays tend to be presented in the filthiest manner and dragged below the level of a cheap burlesque. What the British would never do to their beloved Shakespeare, and what the French would never do to Racine or Molière, the German theater does to classical writers like Goethe, Schiller, Klopstock, or Kleist—apparently on the assumption that vulgarity and shameless displays testify to the "avant-garde" genius of the producer or director.[8] What is most depressing about such performances is the utter contempt for and disregard of the writers' language, the utter unwillingness to allow the text to speak for itself (before being instrumentalized for cheap theatrical effects). This thorough inability to listen might well be compared to an evil spell (one of the long-lasting spells cast by the Nazi regime on German culture). What I appreciated in Handke's play was the author's willingness to listen to language, and also to listen to silence which is the undergirding soil of language. Repeatedly, the flow of conversation in his play is interspersed with pauses and silent retreats—as if to give language a chance to recover and renew itself. It is when emerging out of silence that words regain their ability to sparkle and illuminate (and perhaps to break some evil spells).

To be sure, the theme of Handke's play is very different from that of Schiller's drama. His play revolves around the relation or confrontation between "man" and "woman" (in an almost mythological sense), not the confrontation between civic freedom and tyranny—although one might say that Handke's text also involves a kind of rebellion: against the tyranny of dominant fashion, the tyranny of the media and the abuse of language. In choosing as his title the opening line of Schiller's drama, Handke does not pretend to replicate the classical model or launch a simple renaissance of the classics. As he knows, loyalty to the past cannot mean a mere repetition or reiteration of past formulas. In literature as well as in love, resort to old clichés or stock phrases is already a form of betrayal. In this respect, Handke is a student not only of the classics but also of Nietzsche and his emphasis on revaluation. He is, no doubt, familiar with these famous lines of Zarathustra: "I am of today and before, but there is something in me that is of tomorrow and time to come. I have grown weary of poets, old and new" (once they become clichés). And in his attentiveness to language and the minutest details of observed phenomena, he seems to recall these lines from *Twilight of the Idols*: "One must learn to *see*, one must learn to *think*, one must learn to *speak* and *write*: . . . Learning to *see*—accustoming the eye to calmness, to patience, to letting things come up to it; postponing judgment, learning to go around and grasp each individual case from all sides."[9]

As Nietzsche has taught, loyalty to past memories and to future visions must go together. In no small measure, and hardly unintentionally, Handke's play captures the tenor of Schiller's drama—at least its deeper spirit or promise—and carries it forward into our time. As mentioned before, Schiller's work is a paean to human freedom in its noblest and most elevated form: not to a grasping, selfish or anthropocentric freedom nor to one deriving only from obedience to laws, but to a freedom nurtured by human sensibility in all dimensions. In another context, the poet used the expression "beautiful freedom" (*schöne Freiheit*) to distinguish it from a purely empirical or else "noumenal" capacity.[10] In his play, the embodiment of this freedom is the Marquis Posa, friend of Don Carlos. In the mouth of Posa, Schiller places the most stirring and eloquent words about human freedom—words which, marking the zenith of Western modernity, also redeem the latter of its dismal (colonial) complicities. Under the yoke of abject tyranny, a tyranny aided and abetted by Church and Inquisition, Posa states, humankind has been lowered to a "degraded state" and made "a thing to be played upon." This condition, however, cannot last forever, because humankind is greater than its rulers suspect: "It will burst the chains of a long slumber and reclaim once more its just and hallowed rights." And then, stepping up to King Philip directly, Posa implores him to use his position properly and to remove the spell of domination and mind-control: "The kings of Europe pay homage to the name of Spain; be you the leader of these kings." One word, one sentence from your lips, "one pen-stroke or one motion of your hand can renovate or create anew the earth. Sire, grant us liberty of thought!"[11]

"SINCE WE ARE CONVERSING AND CAN LISTEN TO ONE ANOTHER"

As is well known, Martin Heidegger's later work in large measure revolves around the character and meaning of language, especially poetic language. One of his important postwar writings is titled *One the Way to Language* (Unterwegs zur Sprache). As the title indicates, the book is not an exercise in linguistics; it does not seek to offer a theoretical analysis of language—as if it were possible to speak about language from a standpoint outside of language. Rather, the effort is to find one's way in and through language, by being carefully attentive to its sayings, its cues and intimations. The opening essay in the book is simply called "Language." It begins with these lines: "Man speaks. We speak when we are awake and when dreaming. We always speak, even when we do not utter a single word, but only listen or read." Thus, "we speak continuously in one way or another; and we do so because (as one says) speech is 'natural' to us." According to an old adage, going back to the Greeks—Heidegger reminds as—human beings are said to be "speaking animals," that is, creatures which "have or possess language by nature." However, he adds, one needs to be careful here. The adage does not just mean that, in addition to other faculties, humans also "possess language or the faculty of speech." Rather, the reverse is the case: "Humans are possessed by language or enabled by it to *be* the beings called humans. In the words of Wilhelm von Humboldt, it is as one who speaks that *Mensch ist Mensch*."[12]

At this juncture, however, another point needs to be considered: namely, that language is not simply a fixed structure or a prison without windows. The assumption that humans are "by nature" or "in essence" speaking creatures seems to suggest that language is a static, transtemporal fixture devoid of change or mutation. This assumption is in conflict with the book's title *On the Way to Language* which means that we need to learn from and make an experience with language; differently put: if speaking is distinctly "human," then being "on the way" signals a path of humanization. In following this path, the first thing to discard is the notion of language as a tool or instrument. In Heidegger's words: "What does it mean to speak? The current view declares that speech is the activation of the organs of sounding and hearing. Speech is the communication of human feelings and thoughts." In all of this, speech is simply a human instrument, and to speak effectively simply means to control language more and more thoroughly. Against this view, Heidegger marshals a very different perspective: "We do not [or should not] pounce upon language in order to force it into the grip of preconceived ideas. We do not wish to reduce language to a concept," to a fixture in the arsenal of the *cogito*. This means that we have to become attentive not just to our own feelings, cogitations or machinations, but to language itself, and to accommodate ourselves to the insight (startling at first) that "language speaks" (*die Sprache spricht*). In other words: "To reflect on language demands that we enter into the speaking of language in order to take up our abode in language, that is, within *its* speaking, not merely in our presumptions."[13]

Heidegger is aware, of course, that language can be, and often is, abused or instrumentalized for ulterior purposes. Thus, political and economic elites often degrade language to serve the goals of power and profit; likewise, ideological movements and mainstream media tend to use language as preferred instrument to seduce or brainwash people and to motivate their behavior. For this reason, Heidegger turns to a preeminent form of language, namely poetic language—because poetry (like flute-playing) is intrinsically valuable and commonly does not serve extrinsic or ulterior goals. As one can see, poetry here is granted this preeminent status not for sentimental or purely "aesthetic" motives, but because of its inner integrity and freedom, its unwillingness to be pragmatically useful or manageable.

In the essay cited before, Heidegger turns to a poem by Georg Trakl titled "A Winter Evening" (*Ein Winterabend*). The poem deals with ordinary things and experiences, not with exotic phantasies. As Heidegger says: "Its content is intelligible; there is no word which, taken by itself, is unknown or unclear." In the poem, we encounter snowfall, a window against which the snow falls, the ringing of a church bell, a house, some wandering migrants reaching that house from afar, the interior of the house with a table carrying bread and wine: all quite ordinary things and happenings. Yet, the poem is not just an empirical description of a given situation; nor does it clothe or decorate a description with poetic feelings, sentiments or cogitations. Rather, as Heidegger suggests, the poem "calls" a world into being with manifold things and happenings; it creates a world out of language (similar to Plato's "city in speech"). In calling forth a world, it brings that world close to us, makes it present to us despite its continued absence: "The calling forth calls into a nearness. Yet, its call does not cancel the distance from which it proceeds. The calling calls in itself, here and there—here into presence, there into absence."[14]

In the poem, world and the things and happenings in it are found in a peculiarly intimate nexus where they are neither fused nor torn apart. Heidegger speaks in this context of an "intimacy" (*Innigkeit*) which preserves and respects their "dif-ference" (*Unter-schied*). In his words: "The intimacy of world and things is not a fusion. Intimacy obtains only where the intimate—world and things—differentiates itself cleanly and remains differentiated. In the midst of the two, in the 'between' of world and things, in their *inter*, dif-ference prevails." What we encounter here, on a deep philosophical level, is another formulation of the "ontic-ontological difference" with which Heidegger wrestled all his life and in which the different levels are neither merged nor separated into antagonistic domains. In a somewhat different idiom, the nexus of intimacy and difference can also be expressed in terms of the correlation of "solidarity" and "freedom," where solidarity respects and sustains the freedom of individual agents, just as freedom respects and sustains common solidarity.

In his later work, Heidegger uses another difficult term to designate this constellation: *Ereignis*—a movement where all things and beings are allowed to come into their "own," to achieve their distinct integrity and freedom, without sacrificing their shared being to this integrity. In the cited essay we read: "Difference is not superimposed on world and things as an external relationship. Rather, the difference of world and things integrates (*ereignet*) things into the birth of world, and integrates (*ereignet*) world into the largesse of things." What is important to remember here is that "difference" and "*Ereignis*" cannot simply be fabricated at will; they cannot even be fabricated by words—unless these words emerge from that silence (*Stille*) which alone grants to words their redeeming strength: "Language speaks as the sound of silence. The latter stills by letting world and things 'be.' This letting-be in the mode of quieting is the *Ereignis* of difference."[15]

Another essay in *On the Way of Language* is titled "Language in the Poem" (*Die Sprache im Gedicht*). The essay deals again with Trakl's poetry, especially with a poem called "Springtime of the Soul" which contains, among others, this line: "A stranger is the soul on earth." The line seems to conjure up a Platonic two-world theory opposing to each other sensual and intelligible, temporal and transtemporal realms. This, however, is not Trakl's point. As interpreted by Heidegger (relying on etymology), "stranger" means a sojourner, someone who is "underway" in search of a proper abode. As he says: "The strange soul journeys ahead," not aimlessly, but "in search of the abode where she can stay *as* wandering," that is, can find intimacy with or in difference (*Unterschied*). Thus, the soul does not try to exit or flee from the earth, but rather "seeks" or is in search of it. The soul is in search not in order to dominate or control the earth, but in order to call it forth poetically and thus to "save the earth *as* earth." Trakl's poetry resorts to many diverse formulations to indicate this differential way of being intimately in or with the earth. In one poem we read that the soul as a stranger is "called into undergoing" (*Untergang*). Yet, undergoing here does not mean decaying or expiring, but rather going over into a more recessed, silent, and inobtrusive mode of being. The line in the poem "Springtime" continues with these words: "Spiritually dusk casts its blueness over the dense forest." In another poem called "Summer's End" we read: "Green summer silently slips away, and through silvery night the stranger's steps resound." The noisy busyness of worldly life here settles into quiet care.[16]

In other poems, this change in further underscored by resort to a notion familiar from Meister Eckhart: *Abgeschiedenheit* (apartness or being apart). On several occasions, Trakl calls the sojourning stranger, the one who is "going under" while seeking the earth, "the departed" (*der Abgeschiedene*). However, "departed" here does not mean simply deceased, passed away, or dead; rather, it means being kept apart and thus preserved and salvaged for a new dawn. In Heidegger's words: "The departed is not the deceased in the sense of lifeless. On the contrary: the departed looks ahead into the blueness of spiritual night." In this sense, the departed is "the sojourning stranger, the not-yet-born" (*der Ungeborene*). In this context, Heidegger offers a philosophical elaboration which reveals the profound inspiration of Eckhart. "Apartness," he writes, "dwells as pure, unsullied spirit. It is the emblem of a deeply hidden, quietly radiant blueness which inspires and guides a more silent childhood into the golden promise of a new beginning." Thus, *Abgeschiedenheit* is the herald of a possible *Ereignis* gathering all things together into the intimacy of difference. In this sense, apartness is a gathering spirit, a gathering which "conducts mortal humanity back into a stiller childhood, saving it as a promise for future generations, for a possible rebirth (*Auferstehen*) of humankind in a new dawn." As Heidegger adds, in apartness "the spirit of evil is neither annihilated or denied, nor unleashed or affirmed"; rather, evil is "transformed." However, such transformation is possible only if the human soul remains a stranger or sojourner, maintaining an apartness which "shelters and guides humankind back into a stiller childhood and the dawn of another beginning."[17]

Seen in this light, Trakl's entire poetry is an invitation to sojourning, to participation in the stranger's search for a proper abode or domicile. The invitation is couched in genuine language, in an assembly of words able to retrieve humanity from aimlessness, doom, and decay. In Heidegger's words: "The language of this poetry intimates a transition or crossing-over (*Übergang*). Its path leads from the eclipse of decay over into the twilight blueness of something holy or sacred." Differently put: "Its language sings the song of a homecoming in apartness (*abgeschiedene Heimkehr*), a homecoming which from the lateness of decomposition guides into the dawn of a stiller and still impending beginning." As one should note, Trakl's poetry as interpreted by Heidegger does not just deal with an individual or idiosyncratic experience, but with a broader generational shift—perhaps the rise of a new Axial Age. "The path of the stranger," we read, "leads away from an old degenerate generation. It escorts him to 'go under' into the pristine dawn of a yet-unborn generation." What is the "degeneracy" of the old generation; what is the spell cast over it? Heidegger asks and answers: "The spell or curse of the decomposing humanity is that it is split asunder into the discord among tribes and races. In this situation, each part is unleashed into the turmoil of dispersed and unilateral savagery or bestiality." As Heidegger adds, drawing a broader philosophical lesson: "The curse is not difference as such, but discord (*Zwietracht*)." What needs to happen to dispel this curse is the rise of a new generation "whose difference sojourns ahead from discord into the gentleness of a simple mutuality or entwinement (*einfältige Zwiefalt*)." To accomplish this, the generation has to remain a "stranger" and to follow "the sojourn of the stranger's soul."[18]

NOTES

1. For the "Midnight Song" see "Thus Spoke Zarathustra," Part Four, in Walter Kaufmann, ed., *The Portable Nietzsche* (New York: Viking Press, 1968), 436. The Song was set to music by Gustar Mahler in his Third Symphony, fourth movement.

2. I am relying here on this edition: *The Arabian Nights*, trans. Husain Haddaway, based on the text of the fourteenth-century Syrian manuscript edited by Muhsin Mahdi (New York: Norton & Co., 1990).

3. For this story see *The Arabian Nights*, 11–15.

4. *The Arabian Nights*, 55–65.

5. *The Arabian Nights*, 206–214.

6. *The Arabian Nights*, 7–16, 428. Hölderlin's verses are taken from one of his unfinished poems and reads in German: "*Viel hat erfahren der Mensch/Der Himmlischen viele genannt/Seit ein Gespräch wir sind/Und hören können voneinander.*" See Martin Heidegger, *Erläuterungen zu Hölderlins Dichtung*, ed. Friedrich-Wilhelm von Herrmann (*Gesamtausgabe*, vol. 4; Frankfurt-Main: Klostermann, 1981), 33.

7. See Friedrich Schiller, *Don Carlos*, trans. Frederick W. C. Lieder (New York: Oxford University Press, 1912), Act I, Scene I.

8. I once witnessed the performance of Lessing's *Minna von Barnhelm* by a Hamburg company where actors on the stage engaged in acts of vomiting, urinating, and simulated sex. Unless familiar with the text, an audience might think that the play was meant for the "Reeperbahn" (a notorious amusement district in Hamburg).

9. See Nietzsche, "Thus Spoke Zarathustra," Part Two, and "Twilight of the Idols," in Kaufmann, ed., *The Portable Nietzsche*, 240, 511. In the latter passage, Nietzsche shows himself to be "phenomenologist" *avant la lettre*. Handke, I believe, would also appreciate these lines of Zarathustra (*The Portable Nietzsche*, 188): "Remain faithful to the earth, my brothers, with the power of your virtue. Let your gift-giving love and your knowledge serve the meaning of the earth. Thus I beg and beseech you. Do not let them fly away from earthly things and beat their wings against eternal walls."

10. See Schiller, *On the Aesthetic Education of Man, in a Series of Letters* (English and German facing), ed., and trans. Elizabeth M. Wilkinson and L.A. Willoughby (Oxford: Clarendon Press, 1967); also my "Beautiful Freedom: Schiller on the Aesthetic Education of Humanity," in *In Search of the Good Life: A Pedagogy for Troubled Times* (Lexington, KY: University Press of Kentucky, 2007), 116–137.

11. Schiller, *Don Carlos*, Act 3, Scene 10. Posa's speech before the king contains a number of other jewels, such as these: "Let happiness flow from your horn of plenty. Let man's mind ripen in your vast empire. . . . Devote to your own people's bliss thy kingly power, which has too long enriched the greatness of the throne alone. Restore the prostrate dignity of human nature, and let the subject be, what once he was, the end and object of the monarch's care." Schiller has been taken to task by some literary critics for demanding "liberty of thought" (*Gedankenfreiheit*) rather than liberty of speech—but mistakenly. For, freedom of speech and expression depends on the prior liberty of thought, the ability to *think* freely for oneself—an ability which is squashed by political or clerical dogmatism and mind-control (in our time one must add: by ideologies and media manipulation).

12. Heidegger, "Die Sprache," in *Unterwegs zur Sprache* (Pfulingen: Neske, 1959), 11. For some reason, the English translation of the essay has been placed in a volume titled *Poetry, Language, Thought*, trans. and ed. Albert Hofstadter (New York: Harper & Row, 1971), 189. I have slightly altered the translation for purposes of clarity and emphasis.

13. *Unterwegs zur Sprache*, 12, 14; *Poetry, Language, Thought*, 190, 192.

14. *Unterwegs zur Sprache* 18, 21; *Poetry, Language, Thought*, 195–196, 198–199.

15. *Unterwegs zur Sprache*, 24–25, 30; *Poetry, Language, Thought*, 202–203, 207.

16. Heidegger, "Die Sprache im Gedicht," in *Unterwegs zur Sprache*, 39–43. For an English translation see *On the Way to Language*, trans. Peter D. Hertz (New York: Harper & Row, 1971), 161–164. As one should note, "blueness" for Heidegger and for Trakl is an emblem of the holy or sacred. As Heidegger states (*Unterwegs zur Sprache*, 44; *On the Way to Language*, 166): "Blueness is not a metaphor for the meaning of the sacred. Rather, blueness itself is sacred in virtue of its gathering, revealing-concealing depth." Perhaps one should note here that in Hindu religion—which has preserved many things forgotten or discarded in Western religion—Lord Krishna is commonly portrayed as blue.

17. *Unterwegs zur Sprache*, 52, 55, 66–67; *On the Way to Language*, 172, 175, 185.

18. *Unterwegs zur Sprache*, 50, 74; *On the Way to Language*, 170–171, 191. Readers of the above comments will no doubt be reminded of these words of Zarathustra: "From all mountains I look out for fatherlands and motherlands. But home I found nowhere; a fugitive am I in all cities and a departure at all gates. Strange and a mockery to me are the men of today to whom my heart recently drew me; and I am driven out of fatherlands and motherlands. Thus I now love my *children's land*, yet undiscovered, in the farthest sea: for this I bid my sails to search and search." See *The Portable Nietzsche*, 233. Compare also this statement in "The Wanderer and His Shadow": "Rather perish than hate and fear, and twice rather perish than make oneself hated and feared—this must someday become the highest maxim for every commonwealth too." *The Portable Nietzsche*, 72.

FIVE

Mindfulness and Art

Naturing Nature

Art does not reproduce the visible but makes visible. —Paul Klee

In the conventional view—a view still widely shared today—art is a special or exceptional product. It is special because it is produced by special human beings called "artists"; in the case of great art, the producer is granted the title of "genius." As a special individual, the artist is assumed to have a particularly astute mind capable of discerning characteristic features (such as three-dimensionality) of the external world including nature; in addition, his/her mind is said to be endowed with special "aesthetic" qualities enabling the artist not only to know but to sense or feel external objects. In a derivative sense, people as viewers or consumers of art are said to need a certain "aesthetic" feeling access to the artist's mind and work. It is mainly for this reason that, in modernity, the "theory" of art has come to be known as "aesthetics."

It should be clear, at this point, that such a theory is limited and peculiarly suited to the modern age. In premodern times, art works tended to be anonymous and few people cared about the artist's mind or special feelings. Moreover, art works were not commonly exhibited in galleries but were embedded in public life. Against this background, it can be seen that the art described above is part and parcel of the modern Western worldview, especially the Cartesian worldview separating inside and outside, mind and matter (or nature). In the same way, this art fits into an age of productivity, with the great artist paralleling master engineers and captains of industry.

In the West, the scenario sketched so far has been challenged and tendentially undermined during the last hundred years; struggling to

71

liberate themselves from the Cartesian stranglehold, new art forms and conceptions of art have emerged and gained a foothold, conceptions which have come to be labeled (somewhat misleadingly) "modern art." This struggle or breakthrough can be observed in all art forms, including music, architecture, and poetry; in the following, I shall concentrate, however, on the primordial visual art form: painting. As it happens, the radical changes occurring during the last century have been registered not only by artists and the viewing public, but also by philosophers and art critics. Among philosophers, it was particularly Martin Heidegger who has tried to rescue thinking about art from traditional "aesthetic theory" and to lift it up to a new level of mindfulness; his first major breakthrough in this field came with his lectures on "The Origin of the Work of Art" (of 1935/1936). To be sure, Heidegger's re-thinking of art was not an isolated venture, but was paralleled or supplemented by the work of numerous other thinkers, including Theodor Adorno and Maurice Merleau-Ponty .[1]

What is remarkable is that, at least in the field of painting, the efforts of philosophers have been matched and sometimes outdone by leading painters reflecting on their works. Prominent and exemplary among the latter were Paul Cézanne and Paul Klee. In the following, I shall start from Heidegger's "Art Work" lectures, turn next to some parallel reflections by Merleau - Ponty and Paul Klee, in order finally to comment on the broader significance of "modern art" for our time.

"THE ORIGIN OF THE WORK OF ART"

Heidegger's lectures on the "Art Work" were presented in the mid-1930s, in a context of political and intellectual drama. Politically, it was the time of the consolidation of the Nazi regime. Intellectually, and with particular relevance for present purposes: it was the time of Heidegger's incipient philosophical transformation or "*Kehre*"—away from the "existentialist" focus on *Dasein* in the direction of a more holistic perspective. It was in the context of this "turning" that art and the nature of art works emerged for him as a crucial issue—an issue largely bypassed in his early writings. To be sure, a personal interest in art, especially painting, was present throughout his early years; particularly noteworthy in this respect was his fascination with the paintings of Paul Cézanne.[2] However, this fascination did not directly penetrate the detailed phenomenological discussions in *Being and Time* (1927). As is well known, Heidegger in that text distinguished mainly three modes of being: human *Dasein*, equipment (*Zuhanden-Sein*), and observed objects (*Vorhanden-Sein*). While *Dasein* referred to beings for whom the meaning of their Being is at stake, equipment denoted entities available for everyday "handy" use and "*vorhanden*" objects targeted epistemic knowledge and analysis. As can readily be seen, none of these modes of being were helpful or suitable for the understanding of works of art. In addition to this lacuna, the text also did not make room for the discussion of "things" in their "thinghood" (free from instrumental or cognitive designs).

The text of the "Art Work" lectures opens with an immediate attack on the conventional conception of the artist as the creative source or "origin" of works of art. "On the usual view," Heidegger writes, "the work arises out of and by virtue of the activity of the artist." This view, however, conjures up instantly the question of the origin of the artist. It is obvious that the latter cannot be called an "artist" prior to or independently of the existence of the art work. Hence, artist and art work mutually presuppose each other or stand in a circular relation—which does not augut well for a proper understanding of the art work and its origin. To make some headway in this field, Heidegger chooses an approach which, at first glance, does not seem very promising: an approach which focuses on the thing-character or "thinghood" of the art work itself. One evident advantage of the approach is that it bypasses traditional aesthetics with its reliance on "aesthetic" feelings or experiences unleashed by art; thinghood leaves no room for such self-indulgent subjectivism. In Heidegger's words: "The much-vaunted aesthetic experience cannot get around the thinghood of art works: something 'stony' is in architecture, something wooden in carving, something colored in painting, something sounding in music." Thus, thinghood is undeniably present in art works; in fact it is so "irremovably present" that one can almost say: "The architectural work resides [or finds its grounding] in stone, the carving in wood, the painting in color, the musical composition in sound." Unfortunately, this statement still leaves completely unanswered the question regarding the meaning of "thinghood"—a question conventional aesthetics has assiduously avoided.[3]

According to Heidegger, traditional Western philosophy has tackled the problem of the thing or thinghood in three different ways—none of which ultimately turn out to be very helpful (when applied to art): First, thing is said to be a "substance" (*res*) with attributes; secondly, it is identified with the target of sensation or sensory experience; and finally, it is pinpointed as the material substrate in the correlation of matter and form, something available to be formed. For Heidegger, the definition of thinghood in terms of substance with attributes seems to capture "the natural outlook" and hence is widely popular. Nevertheless, it harbors two main pitfalls: it is applicable to all "beings" (including God) and hence misses the specificity of "things"; and as an abstract concept it does not properly respect things but "assaults" them. The second definition focusing on sensation is also widely popular and as correct and incorrect as the first—but it is incorrect for the opposite reason, namely, by claiming a cozy proximity to things. In terms of the text: "This view implies not so much an assault upon things as the exaggerated attempt to bring them into the greatest possible nearness . . . (Thus) while the first interpretation keeps things at arm's length, the second makes them press too closely upon us. In both cases, thinghood vanishes." The third approach, finally, treating a thing as "formed matter" has a particularly venerable lineage and, in a way, has long served as the model for art theory or aesthetics. Its advantage is that it avoids the polar extremes of the first two; its drawback is that it is again too general to foster the specific understanding of art works. "Form and content [matter]," we read, "are the most hackneyed concepts under which anything and everything can be subsumed. And if form is linked with the rational and matter with the irrational . . . and if this linkage is further coupled with the subject-object scheme, then a conceptual machinery is unleashed that nothing can withstand."[4]

Thus, exploration of the three approaches has ended in failure: "What emerges is that the prevailing thing-concepts obstruct access to the thinghood of things, and by the same token, access to the equipmental character of equipment and all the more to the 'workly' character of the art work." Yet, despite this failure, the outcome is not entirely negative; for what the exploration shows in each case, is that the thinghood of things is very hard to pin down and, in fact, tends to recede or slip from grasp. "The unpretentious character of things," Heidegger notes, "eludes thought most stubbornly. Yet, could it be that precisely this reticence, this calm unhurried quality belongs to the very nature of thinghood?" If this is so, then a new, hopefully more promising path needs to be taken. The new approach charted in the text consists in the differentiated juxtaposition of thinghood, equipment (*Zeug, zuhanden*) and art work. In this juxtaposition, equipment—familiar from the discussion in *Being and* Time— occupies a kind of middle position in the sense that it straddles the other two; moreover, it has the advantage of being readily familiar due to its character as utensil. "Equipment," we read, "has a peculiar intermediate position between thing and work. Thus, following our hunch we first of all seek to explore the equipmental quality of equipment. Perhaps this approach will give us access to the thinghood of things and the work-character of art works."[5]

For Heidegger, equipment has the additional distinctive quality of being the hidden underpinning of the form-matter schema—but an underpinning covered over by the long history of metaphysics. Borrowing from, but also going some steps beyond *Being and Time*, the text discusses again the character of equipment—using for purposes of illustration a pair of working, perhaps peasant shoes (as depicted, for instance, in a Van Gogh painting). Shoes of this kind clearly are not only to be looked at (*vorhanden*), but to be practically used or worn (*zuhanden*). Moreover, as working shoes, they are used for hard labor, perhaps labor in the fields. For this reason, they have to be strong and sturdy. Heidegger in this context employs the term "usefulness" (*Dienlichkeit*), to indicate that the shoes serve a certain purpose, and also "reliability"

(*Verlässlichkeit*) referring to their dependable quality when used. At this point, the text leaves *Being and Time* behind by introducing a new term not familiar from the earlier work: namely, "earth" in its relation to "world." In *Being and Time*, the term "world" had been employed to denote a framework of possible significance, a broad context of possible meaning. With the term "earth," a counter-point is brought to bear, in the sense that the realm of significance is silhouetted against the reticient, withdrawn, meaning-sheltering earthly domain (in many ways similar to the reticence of thinghood). In Heidegger's presentation, by virtue of their inherent quality, the shoes are placed in the interplay or counter-pull of world and earth: "By wearing the shoes, the peasant [or worker] is inserted into the silent appeal of the earth, and due to their reliability he/ she is assured of the peasant's [or worker's] world. World and earth exist for the wearer, and for people in the same position, in and through the shoes as equipment (*Zeug*)."[6]

With this further clarification of equipment, Heidegger feels able to move on to art works—which exhibit similar features but on a different level. Using again the example of working shoes, he asks: What happens to them when they are placed in an art work—say, Van Gogh's painting titled "A Pair of Shoes"? It is clear that the shoes in the painting are not equipment in that they cannot be directly used or worn. So, what does their pictorial depiction convey? Does it seek to draw attention to the artist, the painter Van Gogh?—which would mean a relapse into the metaphysics of subjectivity (the artist as genius). Or does the painting seek to convey empirical information about its content, as for instance the specific individual wearing the shoes or the specific time and locality when and where the wearing occurs?—which would reduce the painting to an epistemic object (*vorhanden*) or source of empirical knowledge.[7] For Heidegger, none of these assumptions reach the level of the art work. Going beyond considerations of artistic subjectivity and the domains of *zuhanden* and *vorhanden* objects, the art work portraying the pair of shoes attempts and achieves something entirely different: namely, the disclosure of the intrinsic meaning or quality of a life of work, its ordeals, joys and sorrows. And this disclosure happens again in the pull and counter-pull of world and earth, of the meaningfulness (inner truth) of a way of life and its reticently sheltered un-meaning. In Heidegger's words, placed in front of Van Gogh's painting (or any great painting), we are suddenly on a different level: "In the nearness of the art work, we are suddenly somewhere else than where we usually tend to be. The art work has disclosed what a pair of shoes *is* in truth." Differently phrased: "Van Gogh's painting is the disclosure of what equipment—here a pair of shoes—*is* or means in truth." In ontological language: "The [ontic] being [of the shoes] emerges into the unconcealedness (*aletheia*) of its Being."[8]

How does this emergence happen? How is disclosure in and through the art work possible? It is certainly not the case that the artist as a gifted individual "makes" the emergence happen or fabricates willfully the disclosure of meaning or truth. Rather, Being simply discloses itself in the art work, just like "nature" discloses or unfolds itself without prompting or compulsion. In this process, the artist surely is involved, but as an engaged participant or midwife. In a sense, the artist's place is in the rhythm of disclosure; in Spinoza's language, the place is on the side of "*natura naturans*," assisting in the "naturing of nature." To resist the impression of a static painting, Heidegger places the art work into verb form, speaking of the "working" of the art work, of a process that puts the art work to work. As he writes: "In the art work the truth of [ontic] beings has set itself to work . . . Thus, a particular being, say a pair of peasant shoes, achieves in the art work stability or standing in light of its Being. Put otherwise: The Being of beings enters into the steadiness of disclosure." Seen in this light, the art work does not offer information about its content or subject matter nor about the character of the artist, but rather is the place enabling the disclosure of truth and meaning of beings. With regard to the meaning of art itself this means: art is "the truth of Being setting itself to work." Still more differently stated (echoing the language of *Being and Time*): The art work is the happening (*Geschehnis*) of truth. What in the art work is put to work is the disclosure of being in its Being.[9]

The last comment, however, needs to be read cautiously and with circumspection. As indicated before, for Heidegger, the art work is also placed in the pull and counter-pull of disclosure and concealment, of world and earth. As he asks: What does the art work put to work, what does it bring into view? His answer: "The work opens up a world. To be a work means to set up a world." Elaborating on this statement the text immediately observes: "World is not the mere collection of countable or uncountable, familiar and unfamiliar things that are 'at hand' (*vorhanden*) . . . It is never an object that stands before us and can be looked at." Rather, the "world worlds"; it is the "ever unobjectifiable context into which we are placed as long as the paths of birth and death, blessing and curse keep us exposed to Being." Wherever world is opened up, all beings and things "acquire their tempo and rest, their farness and nearness, their spaciousness and constraint." On the other hand, earth is what binds every opening or disclosure back into the non-open: "In every opening [of world], earth provides the sheltering abode (*das Bergende*)." This sheltering or withdrawing reticence also happens in the art work. As was stated before: the stone is in the architecture, the wood in the carving, the color in the painting. In setting itself to work, every art work also "works" or works over the non-open. In Heidegger's words: "That into which the work binds itself back and what it discloses as non-disclosing is what we called 'earth.' Earth is what comes forth as sheltering." As he adds, the relation between world and earth should be seen as counterpoint or tension, not (necessarily) as antithesis or dichotomy: "In setting up a world, the art work also sets forth the earth . . . The work moves earth itself into the open region of world and keeps it there. The work lets earth *be* earth." [10]

At another point, the relation between world and earth is portrayed as a "strife" (*Streit*)—a depiction which seems to conjure up dramatic, even tragic connotations. As Heidegger cautions, however, the meaning of the term is distorted if it is equated simply with discord or dispute leading to destruction. Actually, world and earth are linked in "the unity of the art work"—although this unity arises out of mutual contestation. In a genuine contest, Heidegger observes, "the contending parties elevate each other to the level of mutual self-assertion." Such self-assertion, however, does not mean "the rigid insistence on a given state of affairs," but rather "surrender into the hidden distinctiveness" of their mutual belonging. Thus, in strife or contest, each contestant "challenges and carries the other beyond itself." The aim of the contest is not to reach a shallow consensus or an insipid agreement putting an end to the tension. Rather, the point of the art work is precisely to put to work and enact the counterpoint of world and earth. However, because genuine contest achieves its highest point in "the simplicity of intimacy" (*I'm Einfachen der Innigkeit*), the unity of the work arises in the performance or enactment of the contest, displayed in the tensional harmony of the work: "It is in and through the intimacy of strife that the accomplished work finds its inner repose (*Ruhe*)." [11]

Strife or contest in a different though related sense surfaces in the lecture text prominently, especially in its concluding section: namely, under the heading of truth and un-truth. In Heidegger's words: In setting itself to work, and in opening up the contest of world and earth, "the art work discloses the truth (of Being)." As he elaborates, what is involved here is not an epistemic truth, where a statement or assertion matches correctly an empirical state of affairs. Rather, the meaning here is close to the Greek "*aletheia*," a kind of "ontological" truth which discloses the inner significance of beings and how a particular being stands in relation to "Being" as a whole. This relation relies on a certain openness, a "clearing" (*Lichtung*) which permits the discernment of a being in its "truth." Yet, this clearing is not like a maxim of reason which clarifies and explains everything. Rather, the clearing is itself embedded in beings and hence subject to various kinds of concealment. In the words of the text: "The seemingly familiar is at bottom uncanny (*ungeheuer*). The nature of truth, that is, of unconcealment (*aletheia*) is permeated by a refusal (*Verweigerung*). This refusal, however, is not a fault or defect, as if truth denoted an unalloyed unconcealment devoid of hiddennness." For Heidegger, this latter assumption is completely mistaken, because refusal belongs to the very nature of truth; hence one might say: "Truth is by nature un-truth." Yet, the latter statement does not simply mean that truth is falsehood; instead, it indicates a basic counterpoint (similar to the tension between world and earth): "With the notion of sheltering refusal we wish to refer to that counterpoint which obtains in the nature of truth between clearing and concealment. This counterpoint is the source of the original strife."[12]

CÉZANNE AND KLEE

The "Art Work" lectures are by no means Heidegger's only pertinent reflections in this area (although they are frequently treated as such). For one thing, the lectures do not fully explore all the issues raised in the text, for example, the issue of the nature of "things" and thinghood. In this respect, the ensuing years brought a series of writings shedding additional light on the topic, culminating in an essay of 1950 which inserted the "thing" into a broader holistic context (called the "fourfold" comprising earth and sky, mortals and immortals).[13] Still more importantly, Heidegger in later years returned repeatedly to the issue of the status and character of art works. Thus, in the context of a conference held in Athens at the Greek Academy of Arts and Sciences, Heidegger in 1967 presented a lecture titled "The Origin of Art and the Task of Thinking." The lecture counseled a "step back" (*Schritt zurück*): back from the modern worldview with its entanglement in technology and calculating domination. Such a step, Heidegger added, might be a step forward toward "*aletheia*," toward unconcealment embedded in hiddenness. In this sense, the step would also be a move into the domain of art and the origin of the art work. "As a genuine work," we read, "does the art work not point us toward something not disposable, something that shelters itself so that art does not just convey familiar knowledge? In this sense, does the art work not have to respect quietly what shelters itself, what in its self-sheltering fosters in us reticent awe (*Scheu*) in front of something that cannot be planned, calculated, controlled or fabricated?"[14]

As one should add, Heidegger displayed his closeness to art and art works not only in his writings but—more generally and perhaps more deeply—in his conduct and way of life. One of his closest friends, art historian Heinrich Petzet, reports in his recollections: "In my encounters and conversations with Heidegger, art played a significant, even decisive, role from the beginning." Among artists—meaning here mainly contemporary painters—Heidegger had definite preferences. Although he was familiar from early on with impressionist and expressionist paintings, and also with the work of Van Gogh (especially "A Pair of Shoes") as a result of a visit to Holland, Petzet reports that "Cézanne's significance for Heidegger was steadily increasing." He came to know the latter especially through a series of exhibitions in France and Switzerland, exhibitions which also alerted him to Picasso and Braque. However, despite the multitude of encountered works, "Cézanne remained the star toward

which Heidegger's way was moving—the chief witness for a new transformation of art which was being prepared through this master." Somewhat later, he became familiar with a painter who in many ways followed Cézanne by putting art "to work": Paul Klee. In Petzet's words
again: "It was Paul Klee in whom Heidegger finally saw this transformation really accomplished"—a transformation through which art was liberated from conventional "subjectivist" aesthetics and its metaphysical
underpinnings.[15]

Petzet reflects also on some of the possible reasons for these preferences. As it emerges, the latter had something to do—though in a complex way—with the counterpoint of world and earth. While in some of
the works of Van Gogh, the counterpoint was stretched to assume the
character of open strife or antimony, the paintings of Cézanne—especially during his later phase—managed to embed (without erasing) the tension into a relation of mutual belonging (akin to what Heidegger came to
call "*Ereignis*"). As Petzet reports, Heidegger in a conversation once remarked that, with Van Gogh, "the power of 'expression' had exploded all
traditional notions regarding form and color," while in Cézanne "the
saying [meaning] arises from out of matter itself" so that in the art work
"being *as* being" appears. Over the years, Heidegger moved closer and
closer to Cézanne's work, to the point that a physical resemblance
seemed to emerge. Petzet speaks of " the extraordinary intimate way in
which Heidegger was connected to the appearance and being of the
painter." But it was not only in their physical appearance that thinker and
painter were related: "It was their whole manner of being that brought
the two men closer." Later in life, Heidegger frequently visited the places
where Cézanne had worked: the quarry at Bibemus and the way leading
to St. Victoire mountains. During one of these visits—the French thinker
Jean Beaufret reports—Heidegger remarked that Cézanne's route was
"the pathway to which, from its beginning to the end, my own pathway
of thinking responds, in its own way." Petzet refers to another occasion
where Heidegger observed that, in a painting, "the tension of disclosure
and sheltering" can be blended into a "belonging together."[16]

A motivation similar to his close engagement with Cézanne seems to
have been at work in his preference for Paul Klee over many other contemporary painters, including Pablo Picasso. Although several of the latter's paintings "touched him deeply"—Petzet reports—Heidegger
tended to regard Picasso "with a certain 'observer's' distance," perhaps
because he regarded him as an artist-genius (in the older tradition). The
situation with Klee was different. In Petzet's recollection, during the postwar years "the topic of Klee moved rapidly to the center" of his conversations with Heidegger, a fact deriving from the latter's frequent visits to
Klee exhibitions in Basel and elsewhere. Over the years, Heidegger came
to see the painter as "an enormous phenomenon," partly because of his
continuation and distillation of the legacy of Cézanne. As he remarked at

one point, Klee's paintings never intended to portray or represent "objects," but rather tried to capture their coming into being, the way they allow being to "appear." At one point, he even contemplated writing a sequel to "The Origin of the Work of Art," with the focus now shifted to Klee. In 1962, Heidegger delivered one of his final lectures at the University of Freiburg under the title "Time and Being" (a turning-around of the title of his early work). The lecture started with these lines: "If we were to be shown right now two pictures by Paul Klee which he painted in the year of his death—the watercolor 'Saints from a Window' and the tempera on burlap 'Death and Fire'—we should want to stand before them for a long while, and should abandon any claim that they be immediately intelligible."[17]

As previously mentioned, Heidegger was not the only recent philosopher reflecting mindfully on "modern art." With regard to Cézanne and Klee, his observations were ably paralleled by some writings of Merleau-Ponty (whose work he valued and appreciated).[18] One of Mereleau-Ponty's pioneering postwar writings—published roughly at the time of his *Phenomenology of Perception* (1945)—is his essay "Cézanne Doubt." The essay explores the painter's agonizing struggle to find his proper way as a painter. Although surrounded and influenced by impressionist art, he was seeking for something else. When confronted with the impressionist agenda, he is reported to have said: "They created pictures; we are attempting a piece of nature." With these words, Cézanne placed himself far beyond Cartesian metaphysics, with its center in the *ego* and with "nature" located outside of mind. In Merleau-Ponty's words: Cézanne "was never at the center of himself"; rather, it was "*in* the world that he needed to realize his work, with colors upon a canvas." His effort to move beyond the subject-object binary was so radical that he even wanted, as he said, "to confront the sciences with the nature from which they came." In terms of the essay, by overturning traditional metaphysical antinomies—between mind and body, thought and vision—Cézanne "returned to just that primordial experience from which these notions are derived and in which they are inseparable." In a way, the painter struggled to reach access to "*natura naturans.*" As he once stated: "I am oriented toward the intelligence of the *pater omnipotence.*" Interpreting these words, Merleau-Ponty comments: "He was, in a sense, oriented toward the idea or project of an infinite *logos*. . . . Cézanne's difficulties are those of the 'first word.' He considered himself powerless because he was not omnipotent, but wanted nevertheless to portray the world, to change it into a spectacle, to make *visible* how the world *touches* us."[19]

The last writing published shortly before Merleau-Ponty's death in 1961 is an essay titled "Eye and Mind" (composed at the time he was working on his—posthumously published—book *The Visible and the Invisible*). The essay offers probing reflections on the works of both Cézanne and Paul Klee and, over long stretches, on the complex interlacing of visibility and the non-visible (in Heidegger's terms: of disclosure and sheltering). Again, the text is a rebellion against Cartesian mind-body metaphysics. Merleau-Ponty quotes Paul Valéry to the effect that "the painter takes his body with him," and comments: "Indeed, we cannot imagine how a *mind* (*cogito*) could paint. It is by lending his body to the world that the artist changes the world into paintings." At another point he quotes Cézanne saying "Nature is on the inside," and comments: "Quality, light, color, depth, which are there before us, are there only because they awaken an echo in our body and because the body welcomes them." Elaborating on this comment, the text adds: "There is a human body when, between the seeing and the seen, between touching and the touched, between one eye and the other, between hand and hand, a blending of some sort takes place." For Merleau-Ponty, painting "instantiates no other enigma but that of visibility"; but, while painting, the artist practices "a magical theory of vision." He cites some lines attributed to Paul Klee: "In a forest, I have felt many times that it was not I who looked at the forest. Sometimes I felt that the trees were looking at me, were speaking to me . . . I think that the painter must be penetrated by the universe (and not want to penetrate it)." The text also cites art critic Henri Michaux to the effect that "sometimes Klee's colors seem to have been 'born' slowly upon the canvas, to have emanated from some primordial ground, 'exhaled at the right place' like a patina or a mold." And finally it cites Klee himself saying that "the line no longer imitates the visible; it 'renders visible'; it is the blueprint of a genesis of things."[20]

The last lines are a paraphrase of comments found in a celebrated text by Klee himself: his lecture "On Modern Art" delivered in Jena in 1924. In that lecture, Klee defended modern art against charges of "deformation" too often leveled against it by so-called "realists." As he points out, the artist does not grant to external forms of appearance "the compelling significance" they have for realist critics. The artist does not feel bound by these "realities" because he does not see in them "the essence of the creative process of nature"; more important than the culminating or finished forms are "the formative forces." Thus, the artist examines the visible things formed by nature, but he does so "with a penetrating look." And the more deeply he gazes, the more he finds something else going on: "What imprints itself on him, rather than the finished natural image [*natura naturata*] is the image of creation as genesis [*natura naturans*], for the artist the sole essential image." To be sure, the formative processes are not as readily visible as finished objects; they are in a way hidden or sheltered—but they are not nothing. The task of the artist is to bring the hidden to appearance or render it "visible"; differently stated: to "extend the world-creating activity from somewhere back there forward to the here and now, lending genesis duration." As Klee adds, the artist seems to be a "presumptuous fellow" who wishes to move "from modeled image to primordial image." Yet, this is the artist's lot or aspiration: "to press forward, to achieve some sort of proximity to that secret ground by which the primordial law nourishes every development"—to "the womb of nature which holds the key to everything that is."[21]

WORLD AND EARTH

The forays into new, post-Cartesian terrain forged by painters like Cézanne and Klee have found a deep resonance in contemporary philosophy—as shown in the writings of Heidegger and Merleau-Ponty. But the resonance is far more widespread, reaching beyond the European confines. In fact, the past decade has seen a veritable outpouring of interest in these painters, evident in comprehensive exhibitions of their works in many leading museums in the world. In the case of Klee, a major exhibition took place in 2008 at the McMullen Museum in Boston, organized by a team of art historians and philosophers from around the world (with many of the philosophers being distinguished students of Heidegger and Continental philosophy).

The initiative for the exhibition went back mainly to the American philosopher John Sallis who, in connection with the exhibition, collected also a series of writings on Klee's "philosophical vision." In his own contribution to the volume, Sallis concentrated on the issue where and how Klee's paintings "border" on philosophy, especially the work of Heidegger and Merleau-Ponty. In this context, his essay reviewed probingly some of the pertinent writings mentioned above (from the "Art Work" lectures to "Eye and Mind"). With regard to the relation between Heidegger and Klee, Sallis also mentions some possible differences (not highlighted above). Relying on a suggestion of Otto Pöggeler, he notes that Klee's primordial or "cosmic" leanings may not fully concur with Heidegger's conception of "world and earth." Despite this divergence (not further explored), however, he acknowledges "the affinity between their respective analyses," an affinity which allows Klee's work to "border on philosophy."[22]

The volume resulting from the exhibition contains a number of other searching essays shedding additional light on Klee's philosophical affinities. Rather than delving into these texts, however, I want to return to the relation of world and earth (briefly alluded to by Sallis) and explore its broader contemporary implications. In Heidegger's work, the relation between world and earth has a much wider significance than is often assumed. Let us recall some previously cited lines from the "Art Work" lectures: "In setting up a world; the art work also sets forth the earth. . . . The work moves earth into the open region of world . . . (and) lets earth *be* earth." Later passages in the text elaborate further on this relation or rather counterpoint, now with reference to "truth" (*aletheia*). In Heidegger's words: "The counterpoint in the nature of truth involves the meeting of clearing (*Lichtung*) and sheltering. This meeting is the source of the original strife/contest." The counterpoint establishes a kind of mid-point between world and earth; yet, he adds: "World is not simply an opening corresponding to clearing, nor is earth simply a closure corresponding to sheltering or concealment." Rather, there is a mutual conditioning and interpenetration: "World provides the clearing for the paths of the guiding directions of life." But these directions base themselves on "something not mastered, concealed or sheltered." Hence, earth is not simply a closed region but something "which opens itself up in and through self-sheltering." Thus, earth penetrates world, and world grounds itself on earth in the counterpoint of revealing and concealing.[23]

Heidegger at this point raises the question of how and where this counterpoint of truth happens and can be observed. He answers: "It happens in a few essential ways" — one of which is works of art, for example, Van Gogh's painting "A Pair of Shoes" or a Greek temple reaching up to the sky but resting on solid rocks. But such art works are only one example. Other cases where the counterpoint happens are certain kinds of poetry and, possibly, music. To illustrate the counterpoint of truth, Heidegger also mentions philosophical questioning which "as a questioning of Being addresses the latter in its questionability" (*Fragwürdigkeit*). More unexpectedly, the text mentions as a further example great political actions, more specifically the founding and maintenance of political regimes (*staatsgründende Tat*). Such action occurs in the context of the historical fortunes and misfortunes of a country or people, of the "grandeur and tragedy" of major historical events. In the words of the text: "Wherever world opens itself up, it opens to people the historical possibilities of victory and defeat, blessing and curse, domination and servitude. The upsurge of world makes evident the still undecided and unlimited character of potentials and thus also makes clear the hidden need for limit and concrete decisions." However, in the opening up of world, earth comes to assert its demands: "Earth here means that which sustains everything and which withholds and shelters itself in its own law [*nomos*]. While world grants openness and demands concrete options, earth remains by sheltering its own being." The counterpoint is not a "brute rupture" (*Kluft*), but a contest sustained by "intimate belonging-together."[24]

Heidegger's comments on this score encourage one to pursue the counterpoint to our contemporary world: to that particular "worlding of the world" which is usually called "globalization." The chief engines of globalization are usually found in the expansion of markets, media networks, and military strategies. However, it is clear that the globalizing process also brings into view vast horizons of new learning experiences, cultural interactions, and modes of "being-in-the-world."[25] In many ways, the process opens up a world of fascinating prospects and hitherto unsuspected possibilities, while also harboring the danger of unimaginable catastrophes. Partly due to these dangers, the giddy euphoria of globalization is accompanied by a pervasive counter-pull seeking to safeguard the memory of tradition, the sheltering quality of inherited customs and beliefs, and some kind of local rootedness and identity. Sometimes the contest of global and local pulls is stretched to the breaking point: a situation evident in the radical detachment and irresponsibility of global financial and political elites, on the one hand, and the radical fundamentalism and self-enclosure of local ethnic and/or religious communities, on the other. If we follow the spirit of Heidegger's writings, then the task of our time is to recognize this counterpoint but not allow it to degenerate into a "brute rupture"—which, given contemporary weapons technology, could be catastrophic. As exemplified by some accomplished art works, the endeavor should be to bend—without erasing—the different pulls of world and earth into the mindfulness of inner "repose" (*Ruhe*).

NOTES

1. Epigraph is taken from Klee's "Ueber die moderne Kunst" (1924.) My translation. See Theodor W. Adorno, *Ästhetische Theorie*, ed. Gretel Adorno and Rolf Tiedemann, 5th ed. (Frankfurt-Main: Suhrkamp 1981); *Aesthetic Theory*, trans. Christian Lenhardt (London: Routledge & K. Paul, 1984). For Merleau-Ponty's writings see notes 19 and 20 below.

2. As Otto Pöggeler reports, already in October 1928 Heidegger sent a postcard with a picture of Paul Cézanne to his student Karl Löwith. See Pöggeler, *Bild und Technik: Heidegger, Klee und die Moderne Kunst* (Munich: Wilhelm Fink Verlag, 2002), 171.

3. Martin Heidegger, "Der Ursprung des Kunstwerkes," in Holzwege (Frankfurt: Main, 1963), 7–9; "The Origin of the Work of Art," in David F. Krell, ed., *Martin Heidegger: Basic Writings* (New York: Harper & Row, 1977), 149–151.

4. "Der Ursprung," 12–17; "The Origin," 153–158. Heidegger suggests a bit later that the form-matter formula also attacks or "launches an assault upon things." He also links the formula closely with both Christian theology and modern metaphysics: "The inclination to treat the matter-form structure as the foundation of every being receives an additional impulse from the fact that on the basis of religious faith, specifically biblical faith, the totality of all beings is represented as something created, which here means formed or made. . . . The metaphysics of the modern period rests on the form-matter structure devised in the medieval period . . . [and] has become current and self-evident." See "Der Ursprung," 19; "The Origin," 159–160.

5. "Der Ursprung," 20–21; "The Origin," 160–162.

6. "Der Ursprung," 22–23; "The Origin," 162–164. Heidegger suggest that the pair of shoes are those of a peasant, perhaps peasant woman. This is a plausible interpretation, but of course not mandatory.

7. An inordinate amount of ink has been spilled in efforts to pinpoint the precise owner of the shoes and the time and place when and where they were worn. As I see it, these attempts at epistemic accuracy can be left to themselves (*"auf sich beruhen lassen"*). As Heidegger himself says: "From Van Gogh's painting we cannot even tell where these shoes stand. There is nothing surrounding this pair of (peasant) shoes in or to which they might belong – only an undefined space." See "Der Ursprung," 24; "The Origin," 163.

8. "Der Ursprung," 24–25; "The Origin," 164.

9. "Der Ursprung," 25, 27; "The Origin," 164–166. Heidegger is notorious for verbalizing nouns; thus "Being" should be read as be-ing, world as world-ing, language as language-ing, presence as presence-ing. What this means for Heidegger's own work is that this work also should be *put to work* and thus issue in "doing" (something which happens far too rarely).

10. "Der Ursprung," 31, 34–35; "The Origin," 169–172.

11. "Der Ursprung," 36–38; "The Origin," 172–173.

12. "Der Ursprung," 38–41, 43; "The Origin," 173–177. For a further discussion of the "Art Work" lectures see Iain D. Thomson, *Heidegger, Art and Postmodernity* (Cambridge, UK: Cambridge University Press, 2011); Hubert L. Dreyfus, "Heidegger's Ontology of Art," in Dreyfus and M. A. Wrathall, eds., *A Companion to Heidegger* (Oxford: Blackwell, 2005), 407–419; Julian Young, *Heidegger's Philosophy of Art* (Cambridge, UK: Cambridge University Press, 2001); Joseph Kockelmans, *Heidegger on Art and Art Works* (Dordrecht: Nijhoff, 1985); and Friedrich-Wilhelm von Herrmann, *Heideggers Philosophie der Kunst* (Frankfurt-Main: Klostermann, 1980).

13. See Heidegger, "Das Ding," in *Vorträge und Aufsätze*, 3rd ed., vol. 2 (Pfullingen: Neske, 1967), 37–55; "The Thing," in Heidegger, *Poetry, Language, Thought*, trans. Albert Hofstadter (New York: Harper & Row, 1971), 163–182. Compare also Heidegger, *Die Frage nach dem Ding: Zu Kant's Lehre von den tramzendentalen Grundsätzen* (*Geramtausgabe*, vol. 44; Frankfurt-Main: Klostermann, 1984). The latter text goes back to a lecture course in Winter of 1935–1936.

14. Heidegger, "Die Herkunft the Kunst und die Bestimmung des Denkens," in *Denkerfahrungen*, ed. Herrmann Heidegger (Frankfurt-Main: Klostermann, 1983), 147–148.

15. Heinrich Wiegand Petzet, *Encounters and Dialogues with Martin Heidegger, 1929–1976*, trans. Parvis Emad and Kenneth Maly (Chicago: University of Chicago Press, 1993), 135. For unknown reasons, the translators chose to omit the main title of the German original: "*Auf einen Stern zugehen*" (Moving toward a Star).

16. *Encounters and Dialogues*, 140–144. According to Petzet, Heidegger was also fond of and deeply impressed by Rainer Maria Rilke's letters on Cézanne's art, especially the passage where Rilke speaks of the "great turning" in Cézanne's painting: "The single thing, on which everything depends," the point where the work "no longer has any preferences or fastidious over-indulgences" (142). Compare Rilke, *Letters on Cézanne*, ed. Clara Rilke, trans Joel Agee, 2nd ed. (New York, Fromm International Pub, 1986). One can assume that Heidegger was also acquainted with some of Cézanne's writings about his art; some of them are collected in Richard Kendall, ed., *Cézanne by Himself: Drawings, Paintings, Writings* (Boston: Little Brown, 1988).

17. Petzet, *Encounters and Dialogues*, 146–149, 151. Petzet also indicates (146) that Heidegger was familiar with Rilke's letters to Klee. See also Heidegger, *On Time and Being*, trans. Joan Stambaugh (New York: Harper & Row, 1972), 1.

18. According to Pöggeler, Heidegger after the war read Merleau-Ponty's *Phenomenology of Perception* and found an affinity with the French thinker. Merleau-Ponty died in 1961, just a week before a planned visit to Freiburg. See Pöggeler, *Bild and Technik*, 178.

19. Maurice Merleau-Ponty, "Cézanne's Doubt," in *Sense and Non-Sense*, trans. Hubert L. and Patricia A. Dreyfus (Evanston, IL: Northwestern University Press, 1964), 12–14, 16, 19, 25.

20. Maurice Merleau-Ponty, "Eye and Mind," in *The Primacy of Perception and Other Essays*, ed. James M. Edie, trans. Carleton Dallery (Evanston, IL: Northwestern University Press, 1964), 162–164, 166–167, 182–183. Opposition to Cartesian metaphysics is also evident in these comments (170, 186): "A Cartesian does not see *himself* in the mirror; he sees a dummy, an 'outside,' which he has every reason to believe other people see in the same way . . . A Cartesian can believe that the existing world is not visible, that the only light is that of the mind [*cogito*], and that all vision takes place in God."

21. Paul Klee, "On Modern Art," in *Paul Klee: Philosophical Vision: From Nature to Art*, ed. John Sallis (Boston: McMullen Museum of Art, 2012), 13–14.

22. See John Sallis, "Klee's Philosophical Vision," in Sallis, ed., *Paul Klee: Philosophical Vision*, 17–18; also Pöggeler, *Bild und Technik*, 132. Compare in this context also Stephen M. Watson, *Crescent Moon over the Rational: Philosophical Interpretations of Paul Klee* (Stanford, CA: Stanford University Press, 2009).

23. Heidegger, "Der Ursprung," 43–44; "The Origin," 176–177.

24. "Der Ursprung," 44, 50–57; "The Origin," 177 (the English translation omits the last comments).

25. Compare my *Being-in-the-World: Dialogue and Cosmopolis* (Lexington, KY: University of Kentucky Press, 2013). In that book, I followed mainly the lead of Heidegger's *Being and Time* where human *Dasein* is defined as "being-in-the-world." In the light of Heidegger's later writings, this definition needs to be amended as "being-in-the-world-on-earth."

SIX

Mindfulness and History

The Future of the Past

Herkunft aber bleibt stets Zukunft. —Martin Heidegger

In our time, space is shrinking. Expressions like "global village" or "spaceship earth" testify to this shrinkage or contraction. But I would like to ask here, what happens to time or temporality? Does time equally shrink or contract? There appears to be some evidence for this assumption. Time today does indeed seem to shrink into a null-point or zero-point of the present, and this present seems to be shared in the same way around the globe. Everywhere around the globe the present is marked by consumerism, or by the emphasis on more production for the sake of more consumption, and this consumption is to take place *now* or at least in a succession of repeatable "Nows." One is reminded of Friedrich Nietzsche's query in *The Dawn*: "Are you co-conspirators in the current folly of nations, who want above all to produce as much as possible and to be as rich as possible?" [1]

From the perspective of a triumphalist "Now," both past and future vanish from sight. Under the impact of globalization, humankind everywhere—so it seems—is slipping into a kind of global amnesia where the past is considered simply outdated or obsolete (*passé*). In the same way, the future is exorcised by being streamlined into a sequence of repeatable and predictable "Nows." As an antidote to this shrinkage of time, our age witnesses powerful counter-movements aimed at the recovery of the past and the future for political purposes. On the one side—a side variously labeled "fundamentalism," "communalism," or "revivalism"—the past is erected into the "dead hand" (*manus mortua*) strangling the living; on the

87

other side, the living become the instruments—the expendable instruments—of a futuristic agenda or ideological blueprints.

It is against this background that one needs to approach the theme of a possible "future of the past" or "future of tradition(s)."[2] In this formulation, the future cannot simply be a predictable outcome of the past; at the same time, the past (or past tradition) cannot simply be a means to future ends, an exploitable reservoir for present and future managerial designs. In order to grasp properly the meaning of the phrase "future of the past," one has to view time differently: not as a series of Now-points, but as a lived experience, as an intermingling of sedimented temporalities where the past bears traces of the future and the future traces of the past. It is in this sense that Martin Heidegger says "*Herkunft aber bleibt stets Zukunft,*" meaning that the past is always impending, always opening up new possibilities, thus casting its light (or shadow) on the future. The phrase occurs in a famous dialogue, significantly a cross-cultural exchange: the "Dialogue on Language" between Heidegger and a Japanese. In the course of this dialogue, the Japanese refers to Heidegger's Christian upbringing, and especially his early theological studies, saying: "It is obvious that through your upbringing and your studies you are at home in theology in a manner totally different from those who come from the outside and merely pick up a few things through some readings in that field." To which Heidegger responds: "Indeed, without this theological background I should never have come upon the path of thinking. *Herkunft aber bleibt stets Zukunft*: but the past always lies ahead of us, or is impending, or comes to meet us from the future."[3]

In the following I wish to explore this phrase a bit more closely. As should be clear, the phrase can never be exhausted or exhaustively explained—because then it would be grasped "now" and decay into a mere formula. Hence, my comments can only be exploratory, only steps moving back and forth along a way, without aspiring to any finality. I would like to dwell first of all on Heidegger's notion of "time," as he has developed that notion in a sequence of writings, from some early lectures to *Being and Time*. Next, I want to turn to some broadly parallel articulations offered by some of his contemporaries—chiefly Walter Benjamin and Theodor Adorno—in order finally to indicate the relevance of these thoughts in and for our "time," the time of globalization.

HEIDEGGER AND TEMPORALITY

Heidegger's concern with time dates back to his earliest writings, including his doctoral and post-doctoral dissertations. For present purposes, it should suffice to begin with some of his early lectures and lecture courses. In summer of 1924, shortly after having been appointed professor of philosophy at the University of Marburg, Heidegger delivered a lecture there to the faculty of theology. (One needs to recall here that Marburg at the time was the center of Protestant theology in Germany, numbering among its members luminaries like Rudolf Bultmann, Paul Tillich, and others). The title of his lecture: "The Concept of Time" (*Der Begriff der Zeit*). The lecture starts in a somewhat light-hearted vein — but a vein quite appropriate to its context (the faculty of theology). As Heidegger reminds his audience, theologians tend to approach everything, including time, from the angle of God or eternity — which, however, creates a problem. For if, as theologians admit, theology does not really "know" God or the nature of God, then the attempt to unravel the nature of time by reference to God means to explain one unknown (X) by another unknown (Y). On the other hand, if God is equated with "eternity" and eternity is said to be an endless time or a time that never stops, then we still have not learned anything about the nature of "time" which does not stop. Thus, Heidegger comments a but wrily: "If the philosopher asks about time, s/he is determined to understand time from the angle of time itself."[4]

In the main part of his lecture, Heidegger (more seriously) distinguishes between two senses of time: namely, "nature-time" or "mundane time" (*Natur-und Weltzeit*) and lived time. Nature-time is ordinary clock-time or the time of natural science; basically, this time is predicated on measurement, on the quantifiable sequence of "Nows." "How does time present itself to the natural scientist?" Heidegger asks, and responds: "His determination of time has the character of measurement. Such measurement indicates the 'when' and 'how long,' the 'from-when-until-when.' Time is measured by the clock. And a clock is a physical system which (in the absence of external disturbance) continuously repeats the same temporal sequence." By relying on measurements, nature-time fixes a null-point or Now-point located between a "before" and an "after," but it is a Now-point which in itself is undifferentiated and in no way distinguishable from other Now-points — which means that time here is thoroughly uniform or homogenized. In Heidegger's words: "Time is measurable only by being homogeneously constituted. In this account, time is a continuous movement (*Abrollen*) whose phases stand in the relation of earlier and later. What is earlier and later is determined by a 'now' which itself, however, is completely arbitrary and indistinct."[5]

In contrast to nature- or clock-time, lived time operates on a completely different level: that of differentiated or "authentic" human experience. Proceeding from the angle of clock-time, the lecture asks: "What is this 'Now' [of clock-time]? Am I in charge of the Now? Am I that Now myself?" — finding such formulations entirely misleading. Heidegger at this point enters into a discussion of his view of human existence or *Dasein*, a discussion which in many ways anticipates his later, more extensive treatment of the topic in *Being and Time*. Thus, we find here the articulation of existence or *Dasein* as "being-in-the-world," a being which is al-

ways "with others" (*Mit-einander-sein, mit Anderen sein*) and characterized
by "understanding" and "speech" (or language). More importantly, *Dasein* is essentially marked by "care" (*Sorge*), meaning that *Dasein* is that
kind of being which basically "cares" and worries about its manner of
being, and also about its eventual non-being (that is, death or its "being-
toward-death"). Caring about its being and non-being, *Dasein* discovers
itself not as a fixed entity, but as an open-ended possibility or a horizon
of possibilities (terminating in impossibility or the ceasure of further possibilities). As Heidegger states: "As a human life, *Dasein* is primarily
possibility or its manner of being possible (*Möglichsein*)" which also includes the possibility of a "certain and yet indeterminate 'no-longer' (*Vor-
bei*)." From the angle of time this means that "the basic feature of [lived]
time is the future"—the future seen not as a predictable present or a
sequence of predictable "Nows," but as an open-ended horizon moving
forward with lived experience itself.[6]

As should be clear, moving forward to the future here does not signify
a movement in measurable clock-time. Heidegger says provocatively and
poignantly: "[Lived] time does not have the time to measure time." He
does not ignore, of course, that ordinarily humans—all of us—live on a
schedule, a time-table of deadlines and scheduled appointments; trying
to meet this schedule, we are so busy that we are constantly behind time
or running out of time. "By measuring time and living with clock in
hand, "the lecture observes, "our calculating existence constantly mut-
ters: I don't have time." But time is not something we "have" or possess
like an object or a thing (a watch or clock, for example); in a way, time
"has" us. In trying to measure time, all we do is to confine time to the
null-point of "Now," the Now of the present (*Gegenwart*), and the present
is exchangeable and indistinct. In Heidegger's words:

> Ordinary life lives by clock-time, which means: human care returns
> endlessly to the Now, saying: now, from now until then, to the next
> now. . . . If one tries to define time in terms of nature-time, then Now
> becomes the measure (*metron*) for the past and the future. Time here is
> always already interpreted as the present (*Gegenwart*), the past as no-
> longer-present and the future as indefinite not-yet-present. This means
> that the past is gone (*passé*), the future indeterminate. . . . All events roll
> out of an endless future into the unrecoverable past.[7]

What clock-time can never furnish is a proper understanding of the
past and the future. This is so because neither the past nor the future is a
series of indistinguishable Now-points. "The clock," Heidegger says, "indicates the Now. But no clock can ever tell or indicate the future or has
ever told the past." To gain a proper understanding of past and future we
need to take our bearings from lived experience, more particularly from
human existence (*Dasein*) seen as an open-ended possibility. From this
perspective, every present is already impregnated with its future, just as

it harbors or shelters its past. Pithily phrased: *Dasein* does not "have" time (as a property) but rather *is* time or temporal—just as time itself is not clock-time but lived "temporality" (*Zeitlichkeit*). Being temporal means to be part of a story or to be caught up in a "history" or historical narrative. Just like every lived moment, so this story or history opens up a future horizon which, in turn, gives meaning or significance to the present. "Our access to history," the lecture states emphatically, "is grounded in the possibility to understand the present as future-oriented. This is the first principle of hermeneutics." The same orientation also casts a light on the meaning of the past, thereby rendering the past not obsolete but memorable or amenable to retrieval through ever renewed efforts of recollective interpretation. Thus, the past also has a future or gains its meaning through the future or future retrievals. In Heidegger's words:

> The past remains barred from our present, as long as human existence is not viewed as temporal or historical. However, *Dasein* is in itself temporal/historical insofar as it is its own possibility. In its future-orientation, *Dasein* carries with it its past; it returns to or retrieves it as a mode of being . . . . [Differently put:] Being-toward-the-future opens up time, thereby shapes the present and allows us to recover or retrieve the past as a mode of being.[8]

Concisely formulated, the lecture of 1924 contains in a nutshell central features of Heidegger's conception of time or temporality—features which subsequently were fleshed out and further enriched in numerous writings.[9] Here only a few glimpses of this evolving trajectory can be given. During the summer of 1925, Heidegger offered a lecture course at the University of Marburg on the topic "The History of the Concept of Time" (*Die Geschichte des Zeitbegriffs*)—a course which has been published in English under the title *Prolegomena to the History of the Concept of Time*. As both its title and its projected "Outline" indicate, the lecture course was meant to trace the history of the concept of time, by moving specifically from Aristotle via Newton and Kant to Henri Bergson. However, the lecture course remained a torso and did not follow the Outline to its end. Yet, there are some significant passages which can be lifted up here. As before, Heidegger defines time not as clock-time but as story or historical temporality anchored in human existence, more specifically in the orientation of *Dasein* to its future possibilities. As he states, openness to possibility discloses "the future of *Dasein* and with it its past." The dimension in which *Dasein* can achieve itself as future-oriented or "being-ahead-of-itself" (*Sich-vorweg-sein*) is time or temporality. To which Heidegger adds these lapidary sentences:

> Not *"time is"* [as a thing or object], but rather *"Dasein qua time temporalies its own being."* Time is not something which is found outside somewhere as a framework for world events. It is even less something which

whirls away inside in consciousness [contra Bergson and Husserl]. Rather, time is what makes possible the future-orientation (being-ahead-of-itself) of being-in-the-world, that is, what renders possible the being of care (*Sorge*).[10]

Two years later, Heidegger's *magnum opus*, *Being and Time*, was first published. The bulk of the volume deals with the constitutive features of human existence or *Dasein* (which cannot be recapitulated here). The second part of the book, however, turns to the relation between "*Dasein* and Temporality," and it is here that familiar themes of lived time resurface, now in more elaborate from. Again, authentic human *Dasein* is characterized by its future-orientation, which Heidegger now calls "anticipatory resoluteness" (*vorlaufende Entschlossenheit*). Resolutely anticipating the future, *Dasein* is oriented toward its own unique possibility or its "eminent potentiality-of-being (*Möglichsein*)." As Heidegger emphasizes again, future here does not mean a sequence of Now-points which, as a series, would be homogeneous and predictable (and thus no longer a possibility but a determined necessity). Rather, future-orientation means an openness to the unfamiliar or unexpected and the readiness to shoulder or endure the unexpected. In Heidegger's words: "Future here does not mean another Now-point, one that has not yet become actual but will sometime become actual; rather, it means the coming advent/adventure (*Kunft*) in which *Dasein* comes toward or approaches itself in its distinct potentiality of being"—which, in turn, is possible only because *Dasein* is not some "thing" but constantly moves or comes toward itself as unfamiliar horizon (or in the mode of self-becoming through other-becoming). As one should add, anticipation of the future also discloses *Dasein's* indebtedness to its own past—where "past" again is not simply *passé* or a *temps perdu*. Taking over its factual, unpredicted existence, Heidegger says, *Dasein* realizes that its own being is anchored in its past, that is, in its "already having-been" (*Gewesenheit*). By accepting its existence, *Dasein* also accepts its "having-been" as part of its being—a "having-been" to which it can constantly return and which it can recollectively interpret from its future: "*Dasein* can be authentically its past only by being its future."[11]

At the interstices of past and future lies the lived present—a present conceived no longer as an indefinite Now-point, but as a task to be shouldered by existence. Nurtured by the past and illuminated by future possibilities, the present becomes the locus of practical engagement and resolute care. It is in this kind of present, in the "unobstructed encounter" with its given situation, that *Dasein* realizes—or is challenged to realize—its being. Here is a crucial passage:

> Recollecting itself in its future-orientation, resoluteness tackles the present situation. The past arises out of the future, in such a manner that the future-past (or past-future) releases the present out of itself.

We call the combined phenomenon of the future-past releasing the present "temporality" (*Zeitlichkeit*). Only because *Dasein* is temporal in this manner does it have the ability of anticipatory resoluteness realizing its authentic possibility. Thus, temporality reveals itself as the real meaning of authentic care.[12]

WALTER BENJAMIN, THEODOR ADORNO

Instead of pursuing further Heidegger's evolving opus, I would like at this point to glance briefly sideways at some of his contemporaries. In the course of the twentieth century, Heidegger was not alone in reflecting intensively on time and temporality. In many ways, his reflections were paralleled or seconded by other writers of his time, especially by intellectuals associated with the early Frankfurt School. One of these intellectuals was Walter Benjamin who, in 1940, penned his famous "Theses on the Philosophy of History." In articulating his Theses, Benjamin drew at least in part on the teachings of Friedrich Nietzsche, especially on *Untimely Meditations* where Nietzsche had attacked both a nostalgic or escapist antiquarianism and a linear type of progressivisim based on a series of predictable Now-points.

One of Benjamin's Theses (XII) recalled explicitly Nietzsche's statement: "We need history, but not in the way a spoiled loafer in the garden of knowledge needs it" (that is, a loafer who distances history into neutral clock-time). For Benjamin, the point of history and the task of historiography was remembrance of the past—not as *passé* nor for its own sake, but as a lived past; remembrance here means to recall the agonies of the past and the victims of history, and thereby to recover the hopes and redemptive sparks buried in the past (Heidegger would say: the possibilities of being). Hope for the future, in Benjamin's view, could not be actualized through social engineering or ideological designs (based on a predictable future), but only through the retrieval of untapped potentialities and promises. As he wrote in another Thesis (VI): the task of historiography is "to seize hold of a memory as it flashes up at a moment of danger" and thus to continue "fanning the spark of hope in the past" (or the spark of future being in "having been").[13]

In his *Minima Moralia*, written between 1944 and 1947, Theodor Adorno commented with sympathy and appreciation on Benjamin's Theses. What attracted Adorno in the Theses was the emphasis on a buried potentiality that was not governed by clock-time or by the inexorable laws of historical development. As he noted, social-historical reflection today must deal with "cross-grained, opaque, and unassimilated material" which in some sense may be "anachronistic," but which is not simply obsolete or *passé* because, in its own way, it has "outwitted" (and continues to outwit) the "historical dynamic." In their Introduction to *Dialectic of Enlightenment* (of 1947), Adorno and Max Horkheimer struck a similar theme. Countering both antiquarian nostalgia and ideological futurism (which both deny existential temporality), they wrote with some urgency: The point today is that "the Enlightenment must reflect upon itself, if humanity is not to be wholly betrayed. The task to be accomplished is not the conservation [or mummification] of the past, but the redemption of the hopes of the past. Today, however, the past is preserved on in the form of a destruction of the past."[14]

GLOBALIZATION AND TIME

In our age of globalization, the way in which the past is chiefly destroyed is through global amnesia, through the growing homogenization of time or the collapse of time into space. In his Marburg lecture of 1924—composed long before the dawn of globalization (in its present sense)—Heidegger made this nearly prophetic statement: Under the auspices of clock-time, "homogenization means or accomplishes a levelling of time into space, into an absolute presence; it means the tendency to expunge time in favor of the present" (conceived as Now-point or null-point).[15] The other way the past is destroyed today is through its mummification, its erection into an idol to be uncritically worshipped and obeyed. This is what Adorno meant by a "conservation of the past" that extinguishes its redemptive sparks for the living and their future. In either case, there is destruction and violence: either the past is sacrificed to a homogenized present and future or the living are sacrificed to the "dead hand" of the past.

These comments are clearly relevant for non-Western, so-called "developing" societies, and especially for a country like India where the past with its traditions inserts itself so powerfully and pervasively into the present and the future.[16] By "past" here I do not mean a uniform past, but rather a multitude of historical narratives surrounding and informing a multitude of cultural, religious, and social traditions. Clearly, there is in India a Vedic past, a Puranic past, a Muslim past, a British-colonial past, and a number of still other pasts (for example: Jain, Buddhist, Christian, Dalit, tribal). Indians living today have to sort out these pasts and to form a judgment as to which pasts deserve remembrance and retrieval because they contain "redemptive sparks" or hopes for future possibilities of being. This is a difficult and agonizing labor, requiring much individual and collective soul-searching. For example, Indians—many Indians— may no longer be ready or inclined in our time to "worship in the nude." But they still need to be ready to face the question, so marvellously explored by novelist U.K. Ananta Murthy: "Why not worship in the nude?"[17] What is it that still may legitimate such an act? What is it that militates against it or renders it increasingly difficult and perhaps objectionable? Given the insertion of the past into the present and future, such a question cannot simply be avoided. Or rather, the question can be avoided only in two limit cases: on the one hand, by amnesiacs (often secular intellectuals) who have severed their relation to the past (or else have replaced it with a Western historiography); on the other hand, by people (often termed "fundamentalists" or "communalists") who construe the past as a shibboleth able to provide solid marching orders for the present.

By way of conclusion, I would like to return to the "Dialogue on Language" from which these reflections began. In this dialogue, Heidegger describes thinking as a "way" or "path" (*Weg*)—not as a stretch of linear Now-points, but a way on which we move backward and forward (as in a story). As he states: "The lasting element in thinking is the way. And ways of thinking hold within them that mysterious quality that we can walk them forward and backward, and that indeed only the way back can lead us forward." To which the Japanese replies: "Obviously, you do not mean 'forward' in the sense of a linear progress, but . . . I have difficulty in finding the right word." Heidegger: "'For' or 'forward'—into that nearest nearness which we constantly hastily miss or bypass, and which strikes us as strange (*fremd*) each time anew when we catch sight of it."[18]

In the same context, Heidegger describes this moving backward and forward along the way also as the "mutual calling or beckoning (*Einanderrufen*) of past and future" which, in its reciprocity, constitutes genuine presence. Somewhat later he refers to the growing danger of derailment, the danger of losing our path or way—a danger now explicitly linked with globalization or what Heidegger calls the "Europeanization of the earth [or world]." The danger, he says, is growing because of "a process which I would call the complete Europeanization (Westernization) of the earth and humankind." The Japanese in response: "Many people consider this process as the triumphal march of reason. At the end of the eighteenth century, during the French Revolution, was reason not proclaimed a goddess?" Heidegger:

> Indeed. The idolization of that divinity is in fact carried so far that any thinking which rejects the claim of [calculating] reason as not originary is simply maligned today as unreason. . . . The delusion is growing, so that we are no longer able to see how the Europeanization [Westernization] of the earth and humankind erodes every genuine being at the source. It seems that all sources are bound to dry up.[19]

This, however, is not the conclusion of the dialogue. At the very end, Heidegger returns to the interpenetration of past and future, calling it now "the arrival/advent of having-been or what has been (*die Ankunft des Gewesen*)." The Japanese: "But the past passes away, is *passé*—how can it arrive?" Heidegger: "Passing away is something other than having-been

(*Gewesen*)" which is rather the continuing "gathering of what endures (*Versammlung des Währenden*)." Perhaps, in a different idiom, one might call this gathering a "*divan*" or "*satsangha*," that is, the recollection of an enduring promise.[20]

NOTES

1. Walter Kaufman, ed., *The Portable Nietzsche* (New York: Viking Press, 1968), 90.
2. This paper was first presented as the keynote address at a conference on "The Future of Tradition" organized by the Forum for Contemporary Theory in Aurangabad, India, December 16–18, 2000.
3. Martin Heidegger, "A Dialogue on Language," in *On the Way to Language*, trans. Peter D. Hertz (New York: Harper & Row, 1971), 10. Hertz translates the German phrase as "But origin always comes to meet us from the future"; I offer above some alternative renditions. For further elaborations on the phrase see Paola-Ludovica Coriando, ed., "*Herkunft aber bleibt stets Zukunft*": *Martin Heidegger und die Gottesfrage* (Frankfurt-Main: Klostermann, 1998).
4. Heidegger, *Der Begriff der Zeit*, ed. Hartmut Tietjen (Tübingen: Niemeyer, 1989), 5–6.
5. *Der Begriff der Zeit*, 8–9.
6. *Der Begriff der Zeit*, 10–14, 17, 19.
7. *Der Begriff der Zeit*, 20, 22–23.
8. *Der Begriff der Zeit*, 19, 22, 25–26. In the same context (25–26), Heidegger sharply differentiates his view of temporality from the doctrines of historicism and relativism.
9. According to Hans-Georg Gadamer, the 1924 lecture constitutes the "embryo" (*Urform*) of the later *Being and Time*. See Gadamer, "The Marburg Theology," in his *Heidegger's Ways*, trans. John W. Stanley (Albany, NY: State University of New York Press, 1994), 30; also Theodore Kisiel, *The Genesis of Heidegger's Being and Time* (Berkeley: University of California Press, 1993), 315. During 1924, Heidegger also composed a longer treatise on the same topic, which so far has remained unpublished.
10. Heidegger, *History of the Concept of Time: Prolegomena*, trans. Theodor Kisiel (Bloomington, IN: Indiana University Press, 1992), 319–320. For the German original see *Prolegomena zur Geschichte des Zeitbegriffs*, ed. Petra Jaeger (*Gesamtausgabe*, vol. 20; Frankfurt-Main: Klostermann, 1979).
11. Heidegger, *Being and Time*, trans. Joan Stambaugh (Albany, NY: State University of New York Press, 1996), Division Two, par. 65, 299 (translation slightly altered).
12. *Being and Time*, Division Two, par. 65, 300 (translation slightly altered).
13. Walter Benjamin, "Theses on the Philosophy of History," in Hannah Arendt, ed., *Illuminations* (New York: Harcourt, Brace and World, 1968), 255–266 (trans. Harry Zohn). See also Peter Szondi, "Hope in the Past: On Walter Benjamin," *Critical Inquiry*, vol. 4 (1978), 291–506; and Friedrich Nietzsche, *Untimely Meditations*, part 2: *On the Advantage and Disadvantage of History for Life* (1874; Indianapolis: Hackett Publ. 6., 1980).
14. See Theodor W. Adorno, *Minima Moralia*, trans. E.F.N. Jephcott (London: New Left Books, 1974), 151; also Max Horkheimer and Adorno, *Dialectic of Enlightenment*, trans. John Cumming (New York: Seabury, 1972), xv.
15. Heidegger, *Der Begriff der Zeit*, 24.
16. As Anita Desai has written: "In India the past never disappears. It does not even become transformed into a ghost. Concrete, physical, palpable, it is present everywhere. Ruins, monuments litter the streets, hold up traffic, and create strange islands in the modernity of the cities." See her Introduction to Attia Hosain, *Sunlight in a Broken Column* (Delhi: Penguin, 1961), vi.

17. U.K. Ananta Murthy, "Why Not Worship in the Nude? Reflections of a Novelist in His Time," in Fred Dallmayr and G.N. Devy, eds., *Between Tradition and Modernity: India's Search for Identity* (New Delhi: Sage Publ., 1998), 313–325.

18. Heidegger, "A Dialogue on Language," 12.

19. Heidegger, "A Dialogue on Language," 12.

20. "A Dialogue on Language," 12, 15–16, 54. The Persian term *"divan"* means collection, gathering, council; the Sanskrit term *"satsangha"* means truth-gathering or goodness-gathering. In Heidegger's expression *"Versammlung des Währenden"* one should not overlook the connection between *"Während"* (lasting) and *"wahr"* (true). Compare in this context also Gadamer's comments on the meaning of "the classical": "It does not refer to a quality that we ascribe to particular historical phenomena but to a notable mode of being historical: the historical process of preservation (*Bewahrung*) that, through constantly proving itself (*Bewährung*), allows something true (*ein Wahres*) to come into being." See Gadamer, *Truth and Method*, 2nd rev. ed., trans. Joel Weinsheimer and Donald G. Marshall (New York: Crossroad, 1989), 287.

SEVEN

Mindfulness and Cosmopolis

Why Cross-Cultural Studies Now?

Peace is the work of justice. —Isaiah 32:17

In 1937, in the midst of the gathering storm in Europe, Martin Heidegger wrote a short essay titled *"Wege zur Aussprache"* or "Paths toward Dialogue." In that essay, he noted that genuine understanding through dialogue among peoples requires two things above all: first, mindfulness of respective historical and cultural differences, and secondly, the long and patient willingness to listen to and learn from one another. Properly conceived, he wrote, mutual understanding involves "the courage to recognize the distinctness of the other," a courage which also entails the "risk" of mutual contestation and transformation.[1] At the time, Heidegger's observations did not fall on fertile ground. Hence, *encore un effort!*

Our contemporary period of globalization has spawned and continues to spawn many new developments: in business, communications, and military technology. Accompanying these developments, but on a more recessed level, are changes in education and cross-cultural understanding. In the present chapter, I want to focus on one particular initiative in the latter area. During recent decades, a new type of scholarly inquiry has emerged which, for better or worse, is called "comparative political theory" and which is an example of broader cross-cultural studies in our time.[2] In the following, I want to indicate the meaning and significance of this new inquiry, its possible contributions, and why it is necessarily a source of disturbance. For some (perhaps even for many), "comparative political theory" is simply a new wrinkle, a new specialty in the array of specialties inhabiting today's academy. From this angle, each specialty or discipline has its own methodology, its own career pattern, and its canonical texts; moreover, it is clearly demarcated from other disciplines and also from the non-academic world. Seen in this light, the new initiative is just an innocuous scholarly niche offering new career opportunities for young professionals in a highly competitive job market.

I want to show here that this view is wrong. Whatever may be its feasibility in the case of "pure" science and "pure" logic, academic self-enclosure does not work in the humanities and the study of politics (including its theoretical study). Since time immemorial, politics and political inquiry have been ranked among the "practical" branches of the tree of knowledge, and such practice is always engaged in and can never be shielded from the "outside" world. Wittingly or unwittingly, no matter how disguised or camouflaged, political inquiry always finds itself embroiled in a context from which it cannot escape. In the case of political thought, self-enclosure necessarily entails the erection of boundaries along "us-versus-them" or "friend-enemy" lines. By insisting on internal self-sufficiency, political thought thus runs the risk of becoming the accomplice of a particular constellation of political power and a shield or battering rod against other constellations. At this point, high-sounding theoretical formulas are in danger of being transformed into geopolitical manifestoes, academic "canons" into lethal "cannons" employed in a festering "clash of civilizations."

In my view, cross-cultural studies are meant not to foster this clash but to serve as its antidote. How is this possible? How, in particular, can comparative political theory perform this role? How should we properly interpret this phrase? In the following, I shall tackle these questions in three steps. First, I ask: What is political theory or philosophy? How should it be construed to serve as an antidote to clash? Secondly: What is the meaning of "comparative"? Does it just mean a spectatorial exercise, a "view from nowhere" which enables us to assess different phenomena in a neutral fashion? If this is not correct, what else is the meaning of "comparative"? Finally, I want to ponder the relevance of comparative political theory, and of cross-cultural studies in general, for contemporary political life.

POLITICAL THEORY AS PRACTICAL PHILOSOPHY

As the name indicates, comparative political theory is a mode of political theory or philosophy and thus stands in a long and venerable tradition. Although there are many different conceptions of political theory/philosophy, it seems preferable at this point to start from common sense usage: whereby it means a thinking or mindfulness about political life. It is not in the first instance a theorizing about theory or a thinking about thought, but a reflection on politics or political life (which, as I mentioned before, is a form of practice or *praxis*). Theorizing from this angle is itself a practice: namely, the mindful endeavor to shed light on political experience. It arises from and is nurtured by the problems or questions thrown up by political life, and it returns to this nurturing soil by trying to distill its meaning (or non-meaning). As I should emphasize, political theory in this sense is by no means the same as a "politicized" theory which tries to harness thought or reflection in the service of a partisan agenda or ideology.

Not being hitched to partisan goals, political thinking is intrinsically a search for truth—not an apodictic or algorithmic truth, but a truth of life nonetheless. For this reason, it is a sister, even the milk-sister, of philosophy seen as *philo-sophia*. In the latter sense, philosophy—in all its branches—is a search for truth, an unsullied and uncontaminated truth. But what does this mean? Friedrich Nietzsche begins is book *Beyond Good and Evil* with these famous lines: "Supposing that truth is a woman—well, then: is the suspicion not well-founded that all philosophers, insofar as they were dogmatists, have not known how to handle woman?" [3] Actually, at this point, we can go a step beyond Nietzsche: it is not just a supposition or an empty speculation; rather it is commonly acknowledged and accepted that truth is a "woman" or has a feminine character. All the major languages attest to this recognition (except, of course, English which is somewhat impoverished in this regard). All assign a feminine gender to truth and wisdom: in Greek, *aletheia* and *sophia*; in Latin, *veritas* and *sapientia*; in French, *la verité* and *la sagesse*; in German, *die Wahrheit, die Weisheit*. Language here reveals a truth about truth: truth like woman cannot be possessed, owned, dominated or controlled. Like truth, woman cannot be appropriated, domesticated or caged; she can only be cherished and pursued, relentlessly, that is, loved. The term "philosophy" hence means love, not ownership of truth and wisdom. [4]

Marshalling his best rhetorical skills, Nietzsche delights in poking fun at the dogmatic proprietors of truth (and since we are here dealing with political thinking, we are all painfully aware of some sects in academia which claim to be such proprietors or owners). He speaks of "the gruesome earnestness, the clumsy importunity with which [dogmatists] have been in the habit of approaching truth using inept and improper means (*ungeschickte und unschickliche Mittel*)." In a somewhat hopeful vein, Nietzsche adds: "Perhaps the time is very near when we shall again comprehend *how* flimsy the cornerstone has been upon which the dogmatists have hitherto built their sublime and absolute philosophical edifices. Perhaps it was only some popular superstition of time immemorial; perhaps it was some play upon words, some seduction of the grammar." [5] In commenting on Nietzsche's text, French philosopher Jacques Derrida has rightly stressed the unavailability and undomesticated character of truth/woman. As he writes: "Woman (truth) will not be pinned down (*La femme/la verité ne se laisse pas prendre*). In truth, woman/truth will not be pinned down. That which will not be pinned down by truth is, in truth, *feminine.*" In order to distinguish truth/woman from a dogmatic assertion or proposition, he adds, Nietzsche is often compelled to place the term "truth" in quotes or "between the tenter-hooks of quotation marks." What is important is to recognize the apartness of truth/woman from any direct access or grasp: "A woman appeals or seduces from a distance. In fact, distance is the very element of her power."[6]

Derrida's reading, in this respect, is helpful and catches an important point. Unfortunately, seduced by a certain "postmodern" delight in paradox, he proceeds to present truth/woman (or truth *as* woman) as simple dissimulation or mere semblance—apparently unaware that, in doing so, he silently re-instates the dogmatic conception of "truth as such" as non-dissimulation or propositional. "There is no such thing," he states, "as the essence of truth/woman because woman averts, is averted from herself. Out of the depths, endless and unfathomable, she engulfs and distorts all vestiges of essentiality, identity or property." And he goes on to say: "There is no such thing as the truth of woman, and it is because of that abyssal divergence of the truth, because that untruth is 'truth.' Woman is but one name for that untruth (*non-verité*) of truth."[7] As it happens, Derrida's delight in riddles here leads him astray: because "untruth" here functions simply as the denial or negation of (propositional) truth—to which, however, it remains tied (in the mode of negation). As Derrida himself has noted in other places, we have to exit from the binary opposition of affirmation and negation, of manifestness and dissemblance—and hence also of truth and untruth (or non-truth). In a way, the relation is similar to that between "Being" (*Sein*) and "nothingness" (*Nichts*) in Heidegger's philosophy—where the latter is the necessary corollary of the former, or the veil which reveals by sheltering and concealing the mystery of Being.[8]

These reflections are particularly important for political thinking as a practical endeavor. In this endeavor, we are concerned neither with apodictic propositional truth nor with the denial of such truth (styled "untruth"). Rather, we are concerned with "practical" truth, the truth of practical political life; traditional terms for such practical truth are justice and the "good life." So we are concerned with questions like these: Who is doing what to whom? Is it right to wage aggressive, perhaps pre-emptive war? Is it acceptable to engage in genocide, ethnic cleansing, and torture? We really want to *know* about these things, and no amount of "postmodern" subterfuge can or should deter us. If we were not concerned about these matters, there would be no sense to the old maxim that one of the tasks of an upright citizen is "to speak truth to power." On this and related points it is good to consult again the political thinker Hannah Arendt. In her essay "Truth and Politics," Arendt distinguishes clearly between epistemic-rational and what she calls "factual" truth (which I prefer to call practical truth). As she indicates, seekers for factual truth (and sometimes even seekers for epistemic truth) have always faced the backlash of the powerful. "Throughout history," she writes, "the truth-seekers and truth-tellers have been aware of the risks of their business: as long as they did not interfere with the course of the world, they were covered with ridicule" (or disdain); but as soon as they interfered, they were "in danger of their life." For Arendt, the opposite of factual/practical truth is not error, illusion or untutored opinion, but rather "the deliberate falsehood or lie" (something with which we are only too familiar in this age of media manipulation). Despite all the myriad obstructions and obfuscations, however, she valiantly upholds the need to speak truth to power: "Although itself powerless and always defeated in a head-on clash with the powers-that-be, truth possesses a strength of its own. . . . Persuasion and violence can (seemingly) destroy truth, but they cannot replace it."⁹

Speaking truth to power cannot be a one-time affair, but has to be a ceaseless endeavor—to be renewed always in new contexts, with new words and new accents. As mentioned before, philosophy (as mindfulness) is a declaration of love to truth and wisdom—and in such declarations use of stock phrases or clichés is strictly prohibited. Since, in practical-political life, truth takes the name of justice or the good life, political thinking is or has to be a love declaration to justice and *Sittlichkeit*; and here again, resort to routines or stock phrases is inappropriate. Each generation and every thinker have to find their own words and vocabulary. This insight militates against the celebration of a frozen set of texts, the "mindless" idolatry of canonical books treated as the embodiment of perennial truth. The point here is not to dismiss or discard canonical books, especially those of the "classical" past; rather, the peril is their mummification. As Martin Heidegger has constantly reminded us, the task of the reader is to retrieve the "unthought" in the thought of past thinkers, that is, to practice creative interpretation which keeps the animating spirit alive. Adapting a phrase used by Pope Leo XXIII, one might say that political thinking is always in need of renewal (*semper reformanda*).¹⁰

This need for renewal militates against the simple equation of political thought with the history of ideas, to the extent that history coincides with chronology (and time with clock-time). No doubt, political thinking can benefit from a close acquaintance with the historical context of ideas; moreover, historical investigation may lead to some form of textual criticism (perhaps even a "higher criticism"). However, the fact remains that historians only are concerned with what was said or done in a given situation; they do not explore the meaning, cogency or "truth" of historical data. In the terminology of Emmanuel Levinas, historians only fasten their attention on the said or done rather than the "saying" or "doing"— which, in the language of Spinoza, corresponds to a preference for "*natura naturata*" over "*naturans*."¹¹ The need for renewal is also not satisfied by a simple expansion of the traditional canon, that is, by the addition of previously marginalized Western as well as non-Western texts. The danger here is that expansion only leads to a structural modification—the addition of new wings to a canonical museum of thought—instead of opening up new horizons of reflection.

"COMPARATIVE" POLITICAL THINKING

In the preceding, my focus has been on theory or philosophy and I have presented it as a love declaration or existential commitment to truth—which, in practical-political life means a commitment to justice and the good life. But here, an important qualification has to be introduced. In the practical-political domain, such commitment cannot merely be a solitary venture—because political life inevitably involves the interaction of many people, where interaction can take the form of contestation, competition, or cooperation. Although certainly requiring personal mindfulness, practical political life occurs in a context which exceeds the confines of individual intentions. In this respect, the centrality assigned by modern Western thought to "subjectivity" or the "thinking ego" requires correction—and it does not matter whether subjectivity is assigned to individual identity or to group or national identity. In this area, political thinking has to accept the turning toward language and communication which is frequently styled as "linguistic turn." Associated prominently with the names of Wittgenstein, Heidegger, Gadamer and Bakhtin, this turn has placed into the foreground the necessarily "relational" or dialogical character of human thought and conduct. In the political arena, the change has led to the re-discovery and re-invigoration of the "public realm" or "public sphere"—and here again, Arendt's work has been crucial with her stress on the connection of publicity and *vita activa*.[12]

What is happening in our time (and what Arendt could barely foresee) is the globalization of the public sphere, that is, the expansion of the public domain in a global and cross-cultural direction. This expansion is particularly challenging for Western culture and thought which, in the past, have always considered themselves as the apex or culmination of cultural and intellectual life. In Derrida's language, Europe or the West has always viewed itself as the avant-garde or promontory (*le cap*) of civilization, as the standard by which others are measured. Politically, this assumption has entailed such ventures as imperialism and colonialism (ventures which have by no means come to an end). In more subtle was, the West's assumed superiority has translated into the imposition of the West's conceptual arsenal, its metaphysical and analytical framework on the rest of the world. It is this presumed prerogative of the West to "define" the non-West which Edward Said—the intellectual who spoke truth to power relentlessly throughout his life—labeled "Orientalism" seen as a corollary of political domination. The remedy for this asymmetry proposed by Derrida and Said is radical and unsettling: in Derrida's terminology it involves a move toward "another heading" (*l'autre cap*) and even to the "other" of any heading or supremacy. In his way, Said pleaded for an "exit" from Orientalism, an exit enlisting the resources of the "hermeneutics of suspicion" as an antidote to the subterfuges and deceptive rhetoric of ruling elites.[13] In both cases, the turn-about was not an invitation to cultural amnesia, to a simple "destruction" of cultural legacies; rather, destruction—better termed "deconstruction"—served as gateway toward a liberation from obfuscation and manipulation.[14]

In the context of Western political thought, the contagion of Orientalism has surfaced most blatantly in the formula of "developmentalism." During several decades after World War II, American political science—especially its subfield of "comparative politics"—was wedded to this formula, taking shape in a massive research program centered at Princeton University and anchored in the proposition that non-Western societies represented a less mature, less "developed" stage by comparison with the West. In such prominent titles as *The Politics of the Developing Areas, Political Culture and Political Development, Aspects of Political Development,* and *Comparative Politics: A Developmental Approach,* the research program was fleshed out and articulated from ever new angles, with the core concept of "development" basically reflected in two trajectories: pointing on the "systemic" level toward growing differentiation and democratization, and on the "cultural" level toward growing secularization and scientific neutrality.[15] Given the political or geopolitical backdrop of "developmentalism"—the Cold War between America and the Soviet Union (styled as the "Evil Empire")—it is not surprising that the academic agenda was buttressed by a semi-intellectual or ideological super-structure: the rise of a belligerent "neo-conservatism" in the political and geopolitical arena allied with the sway of "neo-liberalism" mainly in economics.

Viewed against this background, what is called "comparative political theory" can have little or nothing in common with "comparative politics" as defined by the Princeton agenda. The term "comparative," as used in the former phrase, certainly cannot mean the prejudicial assessment of non-Western societies and cultures from the presumed height of Western "exceptional" perfection. At the same time, however, comparison cannot involve the simple survey and cataloguing of cultures from a detached analytical standpoint, a "view from nowhere" placed outside or beyond all cultural experiences and language games. To make any headway in this field, a few preconditions of cross-cultural comparison need to be stipulated. As it seems to me, the primary precondition is a certain existential and intellectual openness, that is, a willingness to open oneself to the "non-self" manifest in other customs, other idioms, other practices. This openness is not easy to come by, given the ego-centrism endemic to much of Western modernity. In his book *A Secular Age?* Canadian philosopher Charles Taylor has criticized—and rightly so—the retreat of modern subjectivity into the haven of the "buffered self," that is, a self no longer available to novel experiences in the world. Long before Taylor, Theodor Adorno had bemoaned the tendency of the *ego cogitans* to withdraw into itself and to glance at the world as "through the casemate of a fortress."[16] A second precondition of comparative inquiry follows directly from the first: namely, the willingness of the open self to become seriously engaged with the encountered "non-self"—which means to forego the temptation either to abscond into detached neutrality or to plunge into an empty abyss (of "incommensurable" otherness). In this respect, I can invoke the renowned political thinker Bhikhu Parekh, who—in his pioneering book on multiculturalism—correctly insisted that new understanding and a new vision are generated "not by transcending cultural and other particularities, but through their interplay in the cut and thrust of dialogue."[17]

With regard to the "comparison" involved in comparative political inquiry, I have for some time found it helpful and beneficial to invoke insights articulated by the prominent cross-cultural philosopher and philosopher of religion Raimon Panikkar (of both Indian and Spanish descent). Among his other engagements, Panikkar was active in the so-called "East-West Philosophers' Conferences" and also in meetings of the Society for Asian and Comparative Philosophy (SACP) held at the University of Hawaii. At the end of one of these meetings (in 1984) the latter Society published a volume titled *Interpreting Across Boundaries*, which contains Panikkar's seminal essay "What is Comparative Philosophy Comparing?" The essay offers an instructive review of several different meanings or conceptions of comparison, commenting critically on some of them and endorsing in the end a preferable view. One such conception involves lifting comparison up to the level of an "Archimedean fulcrum outside the contending parties," to the "view from nowhere" mentioned before. For Panikkar, such an approach may yield a transcendental or *a priori* philosophy and perhaps a "meta-philosophy"; but it cannot properly be called "comparative" because it exempts itself from the comparison. In his words, comparative philosophizing "cannot accept a method that reduces all [cultural] visions to the view of one single philosophy."[18] In our present context, these comments can serve as a warning signal cautioning us against efforts to restyle comparative political theory as a "global political theory." Quite apart from the fact that there is no real global language, the label exempts political theorizing precisely from the task of serious comparative engagement, of participating in the "cut and thrust of dialogue." In addition, there is the not negligible danger that "global" theory can readily be coopted or appropriated by a global hegemon or superpower.[19]

Another approach closely related to a priori philosophizing, for Panikkar, is structural linguistic analysis (perhaps along the lines of "generative grammar"). The advantage of this approach lays in the fact that it

tries to take language seriously, its disadvantage in the tendency to reduce language to a logical algebra or algorithm, while bypassing the
diversity of idioms as practiced in ordinary life. The derailment into logical algebra might be avoided by resort to a descriptive phenomenology of
practices; but in this case, the theoretical component tends to be bypassed
or truncated. Panikkar at this point turns to his own preferred option
which he calls a "dialogical" or "imparative" mode of philosophizing.
What is involved here is a movement between positions where, starting
from an initial frame of reference, the inquiry exposes itself to novel or
unfamiliar categories and insights. What we have here, Panikkar states, is
"a philosophical stance that opens itself up to other philosophies [or
frames of reference] and tries to understand them from the initial perspective—though it changes in the process." To open oneself up while
changing in the process means a movement of learning—and this precisely is captured in the term "imparative philosophy," where the adjective
does not derive from the Latin *"imperare"* (to command) but *"imparare"*
(to learn). The expression, he explains, is chosen "in order to stress an
open philosophical attitude ready to learn from whatever philosophical
corner of the world, but without claiming to compare philosophies from
an objective, neutral, and transcendent vantage point." Such a philosophizing admits "that we cannot escape taking a stand somewhere" and
does not pretend "to possess a fulcrum outside time and space and above
any other philosophy."[20]

Seen from this angle, comparative theorizing avoids the lure of an
abstract transcendentalism; but does it therefore lapse into a shallow relativism or pluralism where everything is equally valid and acceptable?
With Panikkar, I would say "no" because of the emphasis on dialogue
and mutual learning. To be sure, the task of comparative understanding—of learning through interpretation—is formidable, especially when
we move from one, relatively familiar framework across boundaries to
unfamiliar frameworks. Panikkar speaks in this context of "diatopical
hermeneutics," to underscore the cross-over character of comparison. As
he writes: "Diatopical hermeneutics is the required method of interpretation when the distance to be overcome is not just a distance within one
single culture or a temporal one, but rather the distance between two or
more cultures, which have independently developed in different spaces
(*topoi*) their own modes of philosophizing and ways of reaching intelligibility." Diatopical inquiry implies that we need to "cross boundaries"
and are challenged to become multilingual and (in a way) bifocal or
perhaps multi-focal. It does not involve "the imposition of one philosophical framework" on others, but rather "the forging of a common
universe of discourse in the very encounter"—which is not a sporadic or
one-time affair but a continuous learning enterprise.[21]

The emphasis in the preceding comments must be placed on "forging," that is, seeking or endeavoring, not necessarily on finding. A com-

mon universe of discourse cannot simply be presupposed nor can it be fabricated or engineered. Intelligibility and comprehension sometimes reach limits; in Panikkar's terms, inquiry at this point "touches the shores of the ineffable and thus of silence." Hence, not everything that is compared is commensurable or fully intelligible—which does not mean that frameworks are utterly incommensurable. In the latter case, the labor of comparison would be pointless. All this highlights the importance as well as the difficulty of translation—which inevitably happens "after Babel." In Panikkar's words: "We speak *a* language, and other people speak other languages we do not know. The primordial language is hidden in our respective languages not as *a* [or another] language, but as language."[22] Thus, in our efforts of comparing and translating we implicitly invoke or rely upon language "as such," a language which we cannot fully know or articulate (what Walter Benjamin calls a "Messianic horizon"). The philosopher and Paul Ricoeur-student Richard Kearney, in introducing his teacher's great late text *On Translation* (*Sur la traduction*), has this to say: "Linguistic hospitality calls us to forgo the lure of omnipotence: the illusion of a total translation which would provide a perfect replica of the original. Short of some kind of abstract symbolic logic or fantasy Esperanto there is no single unitary language. Translation is always *after Babel*." And yet there is hospitality—which means that "the task of translation is an endless one, a work of tireless memory and mourning, of appropriation and disappropriation, of taking up and letting go, of expressing oneself and welcoming others."[23]

CROSS-CULTURAL POLITICAL THINKING TODAY

In some sense, comparative inquiry and diatopical translation have always been important and—to a limited extent—have always been practiced. So, what makes this kind of inquiry specially relevant and even urgent today? The answer is not hard to find: it is the specially dismal and perilous condition of our world today. As indicated before, cross-cultural translation and understanding presuppose a certain openness and hospitality, a generosity of spirit welcoming unfamiliar traditions, cultures, and language games. This hospitality today is in very short supply. Everywhere one looks one finds the upsurge of a narrow "identity politics," of a return to the self-enclosed and "buffered self" bemoaned by Taylor and Adorno. Under the aegis of this identity politics, the world is divided more than ever into "us versus them," into friends and enemies, into the Western promontory or capstone and the rest of societies unable or unwilling to catch up. In a highly militarized and partially nuclearized global scenario, the division takes the shape of a new Cold War and, still more ominously, of seemingly endless "terror wars" pitting against each other the proponents and violent opponents of global "order."

In this context, practitioners of comparative political thinking have a number of options. They may wish to exit from the global scenario by retreating into the safety of academia. Shielded by the ivory tower, they may pursue comparative inquiry as a career objective, as a means to secure professional recognition and advancement in this particular field of studies. In this case, the potentially borderless horizons of comparison are confined in the borders of professional benefit. To be sure, non-academics also may engage in cross-cultural pursuits for personal benefit and enjoyment, for example, by savoring the exotic thrills of alien cultures in the course of (individual or group) excursions in distant lands. In this case, comparison becomes an accessory of the global tourist industry. Probably the most dubious use or abuse of cross-cultural comparison consists in the "embedding" of practitioners in corporate economic and/ or military agendas. As is well known, most large companies today employ small armies of cultural "experts" entrusted with the task of testing consumer behavior abroad—under the motto "cultural knowledge is good for business." Embedding is even more widespread and prominent in military ventures. Here, even larger numbers of professionals—translators, sociologists, anthropologists, political scientists, and psychologists—are inducted or instrumentalized in the pursuit of military strategies and/or clandestine operations. Oftentimes, the initial aspiration of academic professionalism ends up in the secure employment of embedded activities.

In almost all these cases, comparative political thinking is appropriated or colonized for extrinsic purposes. However, there is an intrinsic purpose or "good" in this inquiry (pretty much in the Aristotelian sense); and it is this point I want to highlight here at the end.[24] What is this genuine or intrinsic good? In very general terms, one might call it the "humanization" of practitioners, their opening up to the broad range of human experiences. In more traditional vocabulary, one might speak of educational "formation" (*Bildung*) or, with Johann Gottfried Herder, of the "lifting up to humaneness" (*Emporbildung zur Humanität*). How might such formation happen, and how should practitioners properly proceed? Without derailing into methodological scruples, it seems to me that comparativists need to start from the ground up: from the actual conduct and experiential context of the people they encounter and seek to understand. As a general motto one might say: first see, listen, touch, before reasoning; first make room for the multitude of phenomena, before subjecting them to ready-made categories or cognitive schemata. This does not amount to anti-rationalism; it merely says that thinking arises from and is continuously transformed by experience. Here we can invoke Hannah Arendt again who defined the gist of theorizing as "thinking what we are doing," that is, thinking and trying to understand what is going on.[25] Of course, it is not a novel insight but an old wisdom that understanding follows from conduct. It is part of the general patrimony of peoples around the world that one comes to know one another through meals or the breaking of bread; it is in this breaking of bread that heart and mind are opened.

Unfortunately, there is not much bread-sharing going on in our present world, a world riveted by resentments, ethnic cleansing, and terror wars. Thus, if the previously stated assumption—about thinking and practice—is correct, then comparative political thinking can not even get off the ground unless existing hostilities are removed or at least greatly reduced. This proves again the point that human *Bildung* is not an extrinsic but an intrinsic good of comparative inquiry; differently put: bread-breaking is a condition of possibility of this undertaking. To be sure, once the initial barrier is overcome, large fields of inquiry are open to practitioners: they will have to learn about an array of customs, conventions, and taken-for-granted assumptions, enlisting at this point insights culled from anthropology, ethnomethodology, as well as popular literature and art. Having become acquainted with this cultural matrix, practitioners may then want to venture into (what is often called) "high culture," that is, the realm of important philosophical texts, religious traditions and beliefs, and the basic framework of politics and law. The rub here is that these texts, beliefs and public principles are not necessarily those chosen by the practitioner but those which native speakers generally or predominantly consider salient for a proper understanding of their lives. What follows from the preceding remarks is that all these features need to be approached in an attitude of generosity and sympathy—although there are bound to be limits of sympathy and full understanding. In the judicious words of Charles Taylor, long-established customs or practices have a "presumptive worth" in their favor—which, on valid grounds, maybe critiqued and defeated.[26]

Criticism, however, cannot be the first move, because it always supervenes on perception and reception. Differently and more sharply put: criticism is parasitical on acceptance, like negation is on affirmation (or death on life). The poet Rainer Maria Rilke expressed this point in one of his sonnets where he says: "Praising (*rühmen*)—that is what matters."[27] As you may notice, with these comments I am trying to guide readers out into the open, an arena where things begin to matter. The concern at this point is no longer with professional careers; nor are we talking about a small academic sub-discipline in the ivory tower. Certainly the concern is not with tourism or being "embedded" in alien agendas. Out in the open, we suddenly realize our vulnerability and the vulnerability of our world—a world threatened to an unprecedented degree by catastrophes and disasters. We keenly remember the depths of horror to which humankind can sink and has sunk: Auschwitz, Buchenwald, Hieroshima, Nagasaki. We remember not to wallow in despair, but to wake up to a task or challenge: call it the challenge of global *Bildung*, call it our rising or being "lifted up to humaneness."

Given the danger of global nuclear winter today, I think it particularly needful to remember Hieroshima. There is a "Peace Park" in Hieroshima, a call to life and peace in the mist of immense destruction. On the Internet one can find a video-recording of a Japanese musical performance: the performance by a choir of the final movement of Beethoven's Ninth Symphony. The choir numbers ten thousand people, the men in black, the women in white. Beethoven's last movement, as you recall, puts to music Friedrich Schiller's great "Hymn (or Ode) to Joy." I conclude with just a few lines of this hymn or chorus:

Joy, beautiful spark divine,	*Freude, schöner Götterfunken*
Daughter of Elysium,	*Tochter aus Elysium,*
Here we enter drunk-with-fire,	*Wir betreten feuer-trunken,*
Heavenly, your sanctuary.	*Himmlische, dein Heiligtum.*
Your sweet magic re-unites	*Deine Zauber binden wieder*
What custom's edict has divided.	*Was der Mode Schwert geteilt.*
All humans become brothers/ sisters,	*Alle Menschen werden Brüder,*
Where your gentle wings abide.	*Wo dein sanfter Flügel weilt.*

And then comes the great culminating chorus:

Be embraced, you millions!	*Seid umschlungen, Millionen!*
This kiss to the entire world!	*Diesen Kuss der ganzen Welt!*
Brothers/sisters, above the starry sky	*Brüder, überm Sternenzelt*
Must a loving parent dwell.	*Muss ein lieber Vater wohnen.*

NOTES

1. See Martin Heidegger, "Wege zur Aussprache" in Denkerfahrungen (Frankfurt-Main: Klostermann, 1983), 15–21.

2. This chapter was presented as the keynote address at the inauguration of a new Master's degree program in "comparative political theory" a the School of Oriental and African Studies (SOAS) in London on June 13, 2013.

3. Friedrich Nietzsche, *Beyond Good and Evil*, trans. Marianne Cowan (South Bend, IN: Gateway Editions, 1955), Preface, xi.

4. On this point, Hegel went overboard when he claimed that "philosophy" can shed the connotation of "love of wisdom" in favor of possession. In Martin Heideg-

ger's words: "The thinking that is to come can no longer, as Hegel demanded, set aside the name 'love of wisdom' and become wisdom itself in the form of absolute knowledge." See "Letter on Humanism," in David F. Krell, ed., *Martin Heidegger: Basic Writings* (New York: Harper & Row, 1977), 242.

   5. Nietzsche, *Beyond Good and Evil*, xi.

   6. Jacques Derrida, *Spurs: Nietzsche's Styles/Epérons: Les Styles de Nietzsche*, trans. Barbara Harlow (Chicago: University of Chicago Press, 1979), 49, 55, 57.

   7. Derrida, *Spurs*, 51.

   8. See, for example, Martin Heidegger, "What is Metaphysics?" in Krell, ed., *Martin Heidegger: Basic Writings* 91–172; also my "Heidegger and Zen Buddhism," in *The Other Heidegger* (Ithaca, NY: Cornell University Press, 1993), 200–226. See also chapter 4 above.

   9. Hannah Arendt, "Truth and Politics," in Peter Baehr, ed., *The Portable Hannah Arendt* (New York: Penguin Books, 2000), 547, 562, 570. Regarding the notion of "speaking truth to power," compare also Edward W. Said, *Representations of the Intellectual: The 1993 Reith Lectures* (New York: Pantheon Books, 1994), x, xiii, xvi; and my "Speaking Truth to Power: In Memory of Edward Said," in *Small Wonder: Global Power and Its Discontents* (Lanham, MD: Rowman & Littlefield, 2005), 94–114.

   10. See in this context my "Why the Classics Today? Lessons from Gadamer and de Bary," *In Search of the Good Life: A Pedagogy for Troubled Times* (Lexington, KY: University of Kentucky Press, 2007), 141–153; also Mohammed Arkoun, *The Unthought in Contemporary Islamic Thought* (London: Saqi Essentials, 2002).

   11. See Martin Heidegger, *Der Begriff der Zeit*, ed. Hartmut Tietjen (Tübingen: Niemeyer, 1989); also Heidegger, *History of the Concept of Time: Prologomena*, trans. Theodore Kisiel (Bloomington, IN: Indiana University Press, 1985); Jürgen Habermas, *The Past as Future*, trans. Max Pensky (Lincoln, NB: University of Nebraska Press, 1994); Alexander Stille, *The Future of the Past* (New York: Farrar, Straus and Giroux, 2002); also my "The Future of the Past," *Journal of Contemporary Thought*, vol. 12 (Winter 2000), 5–16. Regarding "said" and "saying" see Emmanuel Levinas, *Otherwise than Being or Beyond Essence*, trans. Alphonso Lingis (Dordrecht: Kluwer Academic, 1991), 37–48.

   12. See Arendt, *The Human Condition: A Study of the Central Dilemmas Facing Modern Man* (Chicago: University of Chicago Press, 1958), 45–53, 155–161; also my "Action in the Public Realm: Arendt Between Past and Future," in *The Promise of Democracy: Political Agency and Transformation* (Albany, NY: State University of New York Press, 2010), 83–97.

   13. On the above see Derrida, *The Other Heading: Reflections on Today's Europe*, trans. Pascale-Anne Brault and Michael B. Naas (Bloomington, IN: Indiana University Press, 1992); and my "The Ambivalence of Europe: Western Culture and Its 'Other,'" in *Dialogue Among Civilizations: Some Exemplary Voices* (New York: Palgrave Macmillan, 2002), 49–65. See also Edward W. Said, *Orientalism* (New York: Vintage Books, 1979), and his *Culture and Imperialism* (New York: Knopf, 1993); and my *Beyond Orientalism: Essays on Cross-Cultural Encounter* (New York: State University of New York Press, 1996). Other terms used especially in America for Orientalism and cultural supremacy are "exceptionalism" and "indispensable nation."

   14. For the distinction between the "hermeneutics of suspicion" and the "hermeneutics of recovery" see Paul Ricoeur, *Freud and Philosophy: An Essay on Interpretation*, trans. Denis Savage (New Haven: Yale University Press, 1970), and his *The Conflict of Interpretations: Essays in Hermeneutics* (Evanston, IL: Northwestern University Press, 1974).

   15. See Gabriel A. Almond and James S. Coleman, eds., *The Politics of the Developing Areas* (Princeton, NY: Princeton University Press, 1960); Lucian W. Pye and Sidney Verba, eds., *Political Culture and Political Development* (Princeton, NJ: Princeton University Press, 1965); Lucian W. Pye, *Aspects of Political Development* (Boston: Little Brown, 1966); Gabriel A. Almond and G. Bingham Powell, *Comparative Politics: A Developmental Approach* (Boston: Little Brown, 1966).

16. See Charles Taylor, *A Secular Age* (Cambridge, MA: Belknap Press of Harvard University Press, 2007), 37–42, 134–142, 262–264, 300–307; also Theodor W. Adorno, *Negative Dialectics,* trans. E. B. Ashten (New York: Seabury Press, 1973), 139–140. (Adorno speaks there of a "peephole metaphysics" where the subject, as "a penalty for its deification," is incarcerated in its selfhood, looking into the world as from a fortress).

17. See Bhikhu Parekh, *Rethinking Multiculturalism: Cultural Diversity and Political Theory* (London: Macmillan, 2000), 341; also my "Multiculturalism and the Good Life: Comments on Bhikhu Parekh," in *In Search of the Good Life,* 237–245. A good and concise formulation of the gist of cross-cultural inquiry has been given by Indologist Wilhelm Halbfass who described it as "an open-minded, methodically rigorous, hermeneutically alert, and existentially committed comparative study of human orientations." See his "India and the Comparative Method," *Philosophy East & West,* vol. 35 (1985), 14.

18. Raimon Panikkar, "What is Comparative Philosophy Comparing?" in Gerald J. Larson and Eliot Deutsch, eds., *Interpreting Across Boundaries: New Essays in Comparative Philosophy* (Princeton, NY: Princeton University Press, 1988), 122–123.

19. Panikkar is quite familiar with this danger. As he states at the opening of his essay: "Comparative studies are fashionable today because they belong to the thrust toward universalization characteristic of Western culture. The West not being able any longer to dominate other people politically, it tries to maintain—most of the time unconsciously—a certain control by striving toward a global picture of the world by means of comparative studies." Panikkar, "What is Comparative Philosophy Comparing?" 116. He could not have anticipated the contemporary global information technology which aims to collect all information about the world in a central (mostly secret) archive.

20. Panikkar, "What is Comparative Philosophy Comparing?" 124, 126–128.

21. Panikkar, "What is Comparative Philosophy Comparing?" 130, 132–133.

22. Panikkar, "What is Comparative Philosophy Comparing?" 130, 132.

23. Richard Kearney, "Introduction: Ricoeur's Philosophy of Translation," in Paul Ricoeur, *On Translation,* trans. Eileen Brennan, (New York: Routledge, 2006), xvii, xx. Compare also my "After Babel: Journeying Toward Cosmopolis," in *Being in the World: Dialogue and Cosmopolis* (Lexington, KY: University of Kentucky Press, 2013), 47–58.

24. I have discussed this qualitative horizon before in my "Comparative Political Theory: What is it Good For?" in Takashi Shogimen and Cary J. Nederman, eds., *Western Political Thought in Dialogue with Asia* (Lanham, MD: Lexington Books, 2008), 13–24. With some chagrin I notice that, in a recent report of the American Academy of Arts and Sciences, the point of the humanities and social sciences is said to contribute to "a vibrant, competitive, and secure nation." See *The Heart of the Matter* (Cambridge, MA: American Academy of Arts and Sciences, 2013), 1.

25. Arendt, *The Human Condition,* 6. Even more emphatically she writes in *Between Past and Future*: "My assumption is that thought itself arises out of incidents of living experience and must remain bound to them as the only guideposts by which to take its bearings." See *Between Past and Future: Six Exercises in Political Thought* (New York: Meridian Books, 1965), 14. The assumption is shared, among others, by Aristotle, John Dewey, Alfred North Whitehead, Martin Heidegger, and Maurice Merleau-Ponty.

26. See Charles Taylor, "The Politics of Recognition," in Amy Gutmann, ed., *Multiculturalism and "The Politics of Recognition,"* (Princeton, NJ: Princeton University Press, 1992), 25–73. As I stated earlier: "The central issue here is whether critique proceeds from a presumed self-righteousness or hegemonic arrogance, or else from a shared engagement and a willingness to engage in a *mutually* transforming learning process." See my "Beyond Monologue: For a Comparative Political theory," *Perspectives on Politics,* vol. 2 (June 2004), 254.

27. Rainer Maria Rilke, "The Sonnets to Orpheus" (I, 7), in *The Selected Poetry of Rainer Maria Rilke,* ed. and trans. Stephen Mitchell (New York: Vintage Books, 1989), 234–235.

Appendix A

Robots and Gestell: Are Humans Becoming Superfluous?

The desert grows. — Friedrich Nietzsche

In a recent Internet article, distinguished economist Paul Craig Roberts wrote about the effect of contemporary globalized capitalism, saying that it has "written off the human race." This outcome, in his view, is due in large part to the perversion of the relation between price and profit stipulated by traditional economic theory. In recent times, this theory has been nullified, at least in the West, by governmental policies which "socialize cost while privatizing profits," the latter evident in the massive bail-outs of banks which supposedly are "too big to fail" — bail-outs shouldered as costs by ordinary people. This socializing of costs and privatizing of profits, according to Roberts, is aggravated and intensified by the offshoring of labor jobs and, more importantly, by the potential replacement of human labor by robots. As reported by "robotic experts" at Harvard University, it is now possible to construct mobile machines "programmed with the logic of termites" to be self-organizing and to complete complex tasks without central direction. Surveying these developments, Roberts reaches this somber conclusion:

> The way the world is organized under a few powerful private interests, this technology will do nothing for humanity. Technology means that humans will no longer be needed in the work force and that emotionless robotic armies will take the place of human armies and have no compunction about destroying the humans on whom they are unleashed.

The article ends with this statement: "Harvard technology will prove to be an enemy of the human race."[1]

Glancing at these lines, the philosophically reflective reader is likely to be reminded of the work of Martin Heidegger, especially his observations on modern technology and its potential development into an automated prison-house. As is well-known, Heidegger's thinking after the great war turned increasingly to the problem of modern technology and its proper or improper relation to human life. One of his most well-known and probing discussions of the topic can be found in a lecture he presented in

115

1953 in Munich under the title "The Question Concerning Technology" ("*Die Frage nach der Technik*").[2] Like the "Being-question" pervading his entire work, the technology-question in Heidegger's treatment is not simply an issue regarding the most suitable and efficient instruments of social "progress." The question for him is not located on the level of what, in *Being and Time*, he called "things-ready-to-hand," that is, utensils or "equipment." The issue is not merely pragmatic in a shallow sense, but philosophical—a point missed by his numerous critics. Adepts of technological progress have frequently tended to portray his comments as the ruminations of a backward "woodsman" from the Black Forest, someone out of step with the times. Recent developments, however, show that his observations were not backward but rather near-prophetic. If we take seriously the warnings of Roberts above, then the technology-question is a question not about techniques we use but about the very future of humanity—about how technology uses and abuses us and perhaps renders us ultimately useless.

The opening of Heidegger's essay immediately introduces the distinction between technical gadgets and the meaning of technology. In his words, technical equipment is "not equivalent to the meaning or nature of technology"—a distinction of levels which surfaces also, say, in a discussion of trees. Clearly the meaning or essence of "tree" (or tree-ness) is not simply another tree placed among a multitude of trees. This distinction of levels has also something to do with human freedom or unfreedom, especially in our technological age. As Heidegger observes, as long as we remain focused on the gadgets of technology, in either a supportive or critical attitude, we are necessarily tied to or embroiled in this domain of gadgetry: "Regardless of whether we passionately affirm or reject it, we remain everywhere slavishly chained to this technology." The chains become even more restrictive whenever we view technology as something "neutral" which we may or may not use at will—an attitude which today is "particularly favored." On the other hand, when we shift our focus to the meaning or essence of technology, a new and liberating perspective is opened up. "Our question concerning technology," the essay states, "moves in that direction and, by raising this question, we want to prepare a free relation to technology. Free is such a relation, however, if it opens up to human existence (*Dasein*) a glimpse into the basic meaning of technology." Once we catch such a glimpse, it becomes possible to experience the world of technological gadgets "in its bounds and limitations."[3]

Returning at this point to the everyday empirical (or ontic) level, Heidegger notes that technology is commonly defined in two ways: as a means to given ends; or as a mode of human fabrication. The two views can be called "instrumental" and "anthropological" definitions. Actually, the two views belong together: "for to posit ends and to procure and utilize the required means to them is a human [fabricating] activity."

Hence, technology in the common sense includes "the manufacture and utilization of equipment, tools and machines; the manufactured and used things themselves; and the goals or ends they serve." But here a set of questions arise: What is an instrument? And what is a means-ends relation? In traditional philosophical vocabulary, the latter relation falls under the rubric of "causation" or "causality." Here the further question instantly emerges: What is causation? Is it simply a situation where two factors impinge upon each other, with one factor pushing or "causing" the other factor to move in a certain direction? Or is there another, deeper meaning—which would relate to the factual pushing as "treeness" relates to existing trees? Going back to the Greek (especially Aristotelian) sources of causality, Heidegger finds that the Greek *"aition"* or *"aitia"* did not at all refer to the modern billiard-ball conception but pointed to a relation of mutual indebtedness (*Verschulden*), a mutually engaged complicity or partnership. What happens in this complicity is a movement not of "causation" but of bringing something forward into the openness of disclosure, out of sheltered concealment. In Heidegger's words, the modes of so-called "casuality" play their role "in the movement of disclosure or bringing-forth (*Hervorbingen*)." Through this movement, "the growth of nature as well as the products of the crafts and arts are brought into appearance . . . Bringing-forth guides what is sheltered or concealed into unconcealment."[4]

Seen from this angle, instead of being a mere means-ends correlation, technology as such is also a mode of bringing-forth or disclosure. However, under the impact of modern science, modern technology twists or transforms "bringing-forth" from a disclosure of meaning into an antagonistic confrontation whereby nature is forced or set upon to yield her secrets. In Heidegger's words: "The disclosure operating in modern technology is a mode of challenging or setting-upon (*Herausfordern*) that puts to nature the demand to supply energy such that it can be extracted and stored (for later purposes)." Thus, a tract of land is no longer cultivated in the rhythm of seasons but rather is "challenged" to produce coal and ore; hence, the earth now shows itself forth as a coal mining district, the soil as a mineral deposit. Even fields are no longer cultivated in the traditional sense but become part of the agricultural industry which, in turn, services a "mechanized food industry." The article gives numerous other examples to show how nature now is set-upon and challenged. One such example is a hydroelectric power plant placed into the Rhine river. The plant "challenges the river to supply hydraulic pressure which sets the turbines in motion which, in turn, moves those machines which generate the electric current for which the entire power station and its network of cables are set up to dispatch electricity." Hence, the Rhine is no longer a river where nature displays its own rhythm; rather, the river now is "dammed up into the power plant." Differently put: "What the river is now, namely, a water-power supplier derives from the operation of the power station which sets upon nature."[5]

Although still a mode of disclosure, modern technology is thus pervaded by an antagonism which aims at regulation and control of nature seen as a resource or "standing reserve" (*Bestand*). But as this control is steadily refined and totalized, Heidegger asks, what happens to the human beings who supposedly are "in charge" of technology? The inevitable result is that human beings are likewise challenged, set upon, and reduced to a mere resource. "To the extent that human beings are themselves challenged and set upon," the article says, "are they not also, even more than nature, part of the 'standing reserve'? There is much evidence for it—witness the fashionable talk about 'human resources' (*Menschenmaterial*), the 'supply' of patients in a clinic, and the like." However, if human beings are themselves challenged or targeted, then technology is clearly not simply a human utensil or instrument—as is commonly assumed—a utensil which could readily be discarded. Rather, what is happening is an entire structure or constellation of challenging, where everything and everyone is simultaneously in the state of being challenged or put upon. Heidegger calls this prevailing constellation "*Ge-stell*," that is, a constellation that cannot be removed by a simple act of will. Modern technology, Heidegger states, is thus "no mere human doing"; rather, the challenging structure prevails upon humans "to regulate or control (*bestellen*) everything as a resource or standing reserve" (including humans themselves). Seeking a proper label for this complex process he adds: "We now name that challenging claim which induces human beings to regulate whatever discloses itself as a standing reserve the *Ge-stell*."[6]

At this point, however, something happens which prevents the prevailing constellation from congealing into a prison-house without exit. What emerges is the connection—previously mentioned—between the technology-question and the Being-question: the fact that humans are able to ask about the meaning of Being and hence also about the meaning of any disclosure or unconcealment, including the disclosure present in modern technology. As Heidegger writes: "Precisely because humans are challenged more basically than are the energies of nature . . . they are never reduced to a mere standing reserve." Rather, whenever human beings "open their eyes and ears, unlock their hearts, and give themselves once to thinking and striving, creating and working," they find themselves carried into the realm of disclosure and unconcealment—a disclosure which is not their fabrication but which "calls upon" and "claims" them. With regard to modern technology, the Being-question (*Seinsfrage*) enables human beings to inquire into the meaning of the structure of *Ge-stell*, that is, to call it into question—which is a path toward releasement or a certain free relation toward it. At this point, Heidegger's essay articulates a notion of "freedom" which differs radically from the modern ideology of anthropocentric liberty. "The essence of freedom," he writes, "is originally not connected with the will or the causality of human willing. Rather, freedom means free-being in the sense of a 'clearing' (*Lichtung*) or disclosure. It is to the happening of disclosure and its inner truth that freedom stands in the closest and most intimate kinship." Driving home this point—and deliberately dislodging the modern shibboleth—the essay adds: "All disclosure derives from freedom, points toward freedom and liberates. The freedom of the free consists neither in unfettered arbitrariness nor in constraint by mere laws. Rather, freedom dwells in that sheltered disclosure in whose clearing all truth is veiled."[7]

Seen in this light, modern technology is neither a mere instrument nor a brute destiny condemning us to either blind submission or blind rebellion. Quite the contrary: "By opening ourselves to the question of the meaning of technology, we find ourselves unexpectedly drawn into the pull of a liberating claim." To be sure, this pull operates only as a possibility—which does not instantly or by itself cancel the status of the *Ge-stell* as a possible prison-house and totalizing structure of domination. In Heidegger's presentation, the disclosure happening in modern technology harbors precisely this latter danger. Placed between these possibilities, he writes, modern technology endangers human beings in their humanity. Differently put: technology's disclosure is "not just any but *the* danger"; erected into the constellation of the *Ge-stell*, it is even "the supreme danger" by pushing human beings "to the brink of an abyss where they themselves are taken as nothing but standing reserve (*Bestand*)." At this point, the *Ge-stell* endangers humans in their relation to themselves and to everything that is; it also banishes or eliminates any other form of disclosure. Above all, "it banishes a disclosure which, in a genuine sense of bringing-forth (*poiesis*) allows beings to come forth into appearance." With this banishment, truth as unconcealment (*aletheia*) also vanishes.[8]

What emerges at this point is the urgent need—not for sheer activism or busy-ness—but for an engaged mindfulness attentive to the Being-question and its liberating appeal. Again, such attentiveness is not just an anthropocentric endeavor or project, but implies responsiveness to Being's disclosure. More importantly, what mindfulness yields is awareness not only of technology's immense danger, but also of its "other" side of possible releasement. In the words of the text: "What threatens human beings are not merely the potentially lethal machines and apparatuses of technology." More importantly: "The rule of the *Ge-stell* threatens to deprive humans of the possibility to experience a more genuine disclosure and a more primordial appeal of truth." However, taken as a disclosure of Being, technology cannot entirely block or prevent the upsurge of every alternative disclosure and of every possible appearance of truth. To this extent, the nature of technology necessarily harbors in itself also "the growth of a saving or redeeming power." To allow this saving grace to grow is precisely the task of a mindfulness which is mindful of the primacy of the Being-question—a primacy which enables humans to inquire into the meaning and truth of Being and thereby gain calm and release- ment. In Heidegger's words: This ability "permits human beings to enter and dwell in the highest dignity of their being—a dignity which consists in guarding and preserving the unconcealment as well as the sheltered concealment of all beings on earth."[9]

Roughly two years after the Munich lecture, Heidegger in 1955 pre- sented another talk in Messkirch under the title "*Gelassenheit*," translated later into English as "Discourse on Thinking." In many ways, the talk continued and deepened his earlier observations on technology and its impact on human life. More sharply than before, the new lecture juxta- posed or opposed to each other two kinds of thinking: calculating reason- ing and mindfulness (*besinnliches Denken*), adding that the former was in the process of obliterating the latter. What is happening today, Heidegger stated, is a growing thoughtlessness or mindlessness which is not even recognized or thought about. Thoughtlessness in this sense is an "uncan- ny visitor" (*unheimlicher Gast*) who "comes and goes everywhere in to- day's world." As a result of this visitation, human beings are increasingly "fugitives"—fugitives from mindful thinking or mindfulness. In large measure, this flight is induced by the character of the contemporary age which is often called the "atomic age" and which for some time now has produced a profound "revolution (*Umwälzung*) in all leading concepts and perspectives." As a consequence of this change, the entire world appears as "an object amenable to the attacks of calculating reason, at- tacks that nothing is believed any longer to resist." Hence nature becomes "a gigantic power station, an energy source for industry and modern technology." In fact, technology shapes the relation of humans to every- thing; "it dominates the entire globe."[10]

As Heidegger emphasizes, the issues raised by technology cannot be resolved by resorting simply to more technical gadgets, nor can they be ignored or wished away. The latter is precluded by the relentlessness of technical "progress." Technological advances, he states, "will move faster and faster and can never be stopped"; in all areas of existence, "human life will be encircled ever more tightly by the forces of technology." In some people, these advances even unleash a kind of technical euphoria. Enveloped in thoughtlessness, such people do not stop to consider that "with technological means an attack is being prepared upon the life and nature of humanity composed with which the explosion of the hydrogen bomb means little." The only way to break this bad spell is to remember the basic and most distinctive human quality: the capacity to raise the Being-question and ponder mindfully the meaning of human life. The only promising path on which human beings in the atomic age can pos- sibly make a dent in the overpowering armor of the *Ge-stell* is to marshall and bring to bear mindfulness as an antidote to calculating reason. Mind- fulness today must move beyond the approval or condemnation of techniques and ponder what seems impossible given the power of technology: the potential of a free and non-submissive relation to the atomic age. This is the "other" path open to us: "We can use technical gadgets and yet simultaneously keep ourselves free of them, such that we may let go of them at any time." We can utilize the devices but in the same breath "let them be." Thus, "we can say 'yes' to the unavoidable use of gadgets, and at the same time say 'no' by denying them the right to dominate us and so to warp, confuse and lay waste to our being."[11]

Choosing this option does not involve a simple compromise or half-heartedness—because the path is arduous and requires a resolute break with conformism. Still, the promise of freedom held out by the path is enticing and eminently worth the risks. As Heidegger writes, in following the path "our relation to technology becomes wonderfully simple and calm. We allow technical devices to enter our daily life and at the same time we leave them outside, that is, we leave them be as things which are nothing absolute." This comportment toward technology, he adds, "which comprises a 'yes' and 'no' to technical devices, we can call by an old word *releasement toward things (Gelassenheit zu den Dingen)*." By way of conclusion, however, Heidegger reminds his audience of the extreme precariousness of our condition, of the reigning thoughtlessness entailing the "fugitive" character of mindfulness. The spell exerted by modern technology has by no means been broken. If it persists, the relentless tide of technological revolution could turn out to be "so captivating, bewitching and beguiling that someday calculation will come to be accepted and practiced as *the only way* of thinking." In that case, the successful cleverness of calculating planning would coincide with "indifference toward mindfulness and with total thoughtlessness." At that point, humanity would have "thrown away its most distinctive quality, namely, that human beings are thoughtful or mindful of their being." For this reason, the issue today is "how to save the humanity of human beings."[12]

NOTES

1. Epigraph taken from Nietzsche, *Also Sprach Zarathustra*, Part 4 (1892). My translation. Paul Craig Roberts, "Global Capitalism has Written Off the Human Race," blog/economics/19010–global-capitalism-has-written-off-the-human-race, February 18, 2014. As Roberts adds: "Faced with little demand for human labor, thinkers predict that that the rich intend to annihilate the human race and live in an uncrowded environment served by robots. If this story has not been written as science fiction, someone should get on the job before it becomes ordinary reality."

2. See Martin Heidegger, "Die Frage nach der Technik," in *Vorträge und Aufsätze*, Part I, 3rd ed. (Pfullingen: Neske, 1967), 5–36; "The Question Concerning Technology," in *Martin Heidegger: Basic Writings*, ed. David F. Krell (New York: Harper & Row, 1977), 287–317.

3. "Die Frage nach der Technik," 5; "The Question concerning Technology," 287–288. (In the above and subsequent citations, the translation has been altered for the sake of clarity.)

4. "Die Frage nach der Technik," 6–8, 10–11; "The Question Concerning Technology," 288–290, 292–293.

5. "Die Frage nach der Technik," 14–15; "The Question Concerning Technology," 296–297.

6. "Die Frage nach der Technik," 16–19; "The Question Concerning Technology," 298–301. The term is frequently translated as "Enframing." The German word is formed in this way: One German term for challenging is *"stellen"* — as when a hunter aims at (*stellt*) a deer. A constellation of such *"stellen"* may be called a *"Gestell"* (after the model of *"bergen"* and *Gebirg"*).

7. "Die Frage nach der Technik," 18, 24–25; "The Question Concerning Technology," 299–300, 306.

8. "Die Frage nach der Technik," 18, 24–25; "The Question Concerning Technology," 307–309. In a similar way, but perhaps even more cruelly, animals today are set upon, imprisoned, and enslaved by the operation of the *Ge-stell.*

9. "Die Frage nach der Technik," 28, 32; "The Question Concerning Technology," 309–310, 313.

10. Heidegger, *Gelassenheit* (Pfullingen: Neske, 1959), 11–13, 16–18; *Discourse on Thinking,* trans. John M. Anderson and E. Hans Freund (New York: Harper & Row, 1966), 45–46, 49–50 (in the above citations I have occasionally changed the translation for the purpose of clarity).

11. *Gelassenheit,* 19–23; *Discourse on Thinking,* 51–54.

12. *Gelassenheit,* 23, 25; *Discourse on Thinking,* 54, 56. The gravity of the situation is clearly evident in the case of a young generation totally "bewitched" and "beguiled" by technical gadgets and video games — a thoughtlessness permitting ruling powers to rule absolutely and without disturbance.

Appendix B

Orthopraxis: For an Apophatic Humanism

The Indian social theorist Ananta Giri has asked me to write some after-thoughts to his recent book titled *Knowledge and Human Liberation: Towards Planetary Realization*.[1] The title contains signature notions of the author. In a string of earlier publications Giri has provided sketches adumbrating these notions. Now, in this new volume, he finally has enlisted all his rich intellectual resources in order to flesh out the meaning of an active kind of "knowledge" capable of promoting human "liberation" or emancipation. As one should note, liberation for Giri does not just denote deliverance from oppressive social and political structures, but also the overcoming of inner compulsions and addictions standing in the way of genuine human freedom. To this extent, liberation as used in this volume resonates with Gandhi's notion of self-rule (*swaraj*) and also, more distantly, with the notion of self-care (*cura sui*) as cultivated by the Roman Stoics.

Still, one may ask: a transformative and liberating "knowledge"? Does knowledge not always precede transformative praxis? Does one not first have to know what it is that one wishes to change or liberate? As one can readily see, these questions lead back to an old philosophical conundrum that one can trace from the Greeks (especially Aristotle) down to Hegel and Marx: the so-called "theory-praxis" nexus. Different philosophical traditions give different answers to this conundrum. In large measure, modern Western rationalism—from Descartes to Kant—places the accent squarely on knowledge or cognition, even a purely "transcendental" cognition entirely removed from practical or experiential contexts. On the other hand, Karl Marx is famous for having assigned primacy to transformative social change over cognitive or theoretical interpretation. What seems to be lacking in such extreme formulations is attention to interconnections, to the mutual interpenetration of theory and praxis, of lived contextual experience and cognitive detachment from context.[2] Surely it seems plausible to assume that theoretical mindfulness is triggered by some dilemmas or traumas experienced in ordinary life—and that without this instigation theorizing would be empty and pointless. By the same token, having been triggered by lived dilemmas, reflection can be assumed to make a contribution to the overcoming of these dilemmas, thus paving the way to liberation. There is no doubt this exit route is normally difficult and painful. The Greeks spoke of learning "the hard way" or through suffering (*pathei mathos*).

The close linkage between knowledge and transformative praxis was a hallmark of John Dewey's pragmatism, whose legacy Giri invokes at the beginning of his book. This legacy, it is true, has often been thoroughly misunderstood. On the one side, some philosophers have tried to blend pragmatism into traditional modes of cognition, and especially into conventional forms of epistemology (not far removed from Kantian epistemology, as outlined in the *First Critique*). Profiled against this "pure" theory of knowledge, practice appears as a derivative byproduct of cognition and hence not in any way constitutive of knowledge. On the other hand, pragmatism is notorious for allegedly having equated knowing with doing and even "truth" with "what works." On this construal, pragmatism slides inevitably into consequentialism and even a crude type of relativism—because the criteria for "what works" are not specified and hence are subjectively up for grabs. What is completely neglected in this reading is that "truth" is not a byproduct of work, but is inherently a "work" itself. Differently phrased: truth is not found by simply gazing at the stars, but by human searchers allowing themselves to be "worked over" and thus undergoing a transformative learning experience.

The interlacing of knowledge and praxis is not an exclusive Deweyan insight; it can be found also in Indian thought, especially in the teachings of the Upanishads and the *Bhagavad Gita*. As Giri points out, the *Gita* is well known for its effort to link together three pathways to liberation (or *moksha*): *jñana*, *karma*, and *bhakti*—usually translated as knowledge, action, and devotion. Although making ample room for *jñana* or knowledge, the *Gita* does not allow cognition to overshadow or overwhelm action and devotion. Above all, action is not simply a derivative or marginal byproduct of knowledge; rather, genuine knowledge is made manifest—and intrinsically or primordially made manifest—in right action. By the same token, faithful devotion is not dependent on intellectual insight or knowledge; nor is it seen as antithetical to, or basically at odds with, knowledge. To this extent, the *Gita* makes an important contribution to the resolution of profound dilemmas which have beleaguered Western thought: the compatibility or incompatibility between reason and faith, between philosophy and theology, between immanence and transcendence.

Closely linked with these dilemmas in Western thought has been the issue of the relation between orthodoxy and *orthopraxis*. The issue has gained particular prominence in recent times due to the upsurge of such perspectives as "political theology" and "liberation theology." In the opinion of some critical observers, the linkage of theology with political praxis and liberation threatens the integrity and absolute truth quality of theological doctrines, that is, their "perennial" status outside of practical-temporal contexts. For defenders of these perspectives, by contrast, the removal of "God-talk" beyond practical human reach means to erect an abstract myth devoid of any redeeming and liberating quality; in fact, it is only through participating in God's ongoing self-disclosure—modeled by the image of the "suffering servant"—that religion can shield itself against idolatry and recover the living spirit of faith. To be sure, for defenders, not every type of activism is equally redeeming or liberating, but only a praxis illuminated by its orientation toward goodness and divine grace. In the words of Gustavo Gutierrez, one of the "founders" of liberation theology: "The starting point for all theology is to be found in the *act* of faith. Rather than being a mere intellectual [or theoretical] adherence to the message, it should be a vital embracing of the gift of the Word as heard in the ecclesial community, as an encounter with God, and as love of one's brother and sister. It is about existence in its totality."[3]

As it happens, the relation between orthodoxy and *orthopraxis* has in recent times attracted the attention of numerous theologians and philosophers of religion. A prominent example is Clodovis Boff's book titled *Theology and Praxis: Epistemological Foundations*. In a chapter devoted to the "dialectic of theory and praxis," Boff explores and indicates precisely the mutual interpenetration of theological doctrine and liberating practice. Commenting on this text, one interpreter remarks pointedly that, for Boff, "human practice and involvement" is a basic "constituent of epistemology" or epistemic knowledge; at the same time, his dialectical method "potentially enables the integration of human practice into theoretical understanding."[4] By rejecting the need for full prior knowledge as a requisite for praxis, Boff's text come close to the tradition of "apophatic" theology—a tradition to which Giri refers when he quotes Felix Wilfred's statements about the "ineffable character" of our "human-divine existence" and the "darkness enveloping it." In turn, this tradition is close to the notion of an "apophatic anthropology" as articulated by Ivan Illich in these words: "Apophatic anthropology is the rigor of not talking about God, but actually living as Christ enfleshed has done."[5]

To conclude, let me add these statements by Denys Turner which I find particularly illuminating on the issue of knowledge and praxis:

> At the heart of any authentic spirituality is the means of its own self-critique, an apophatic putting into question of every possibility of "knowing" who God is, even the God we pray to. In the heart of every Christian faith and prayer [probably every faith and prayer] there is, as it were, a desolation, a sense of bewilderment and deprivation, even panic, at the loss of every familiar sign of God, at the requirement to "unknow" God—as the Meister Eckhart put it, for the sake of the "God beyond God." For it is somewhere within that desolation and negativity that the nexus is to be found which binds together the Christian rediscovery of justice with the poor and the rediscovery of the God who demands that justice. For in that bond of action and experience—

"praxis" — is the discovery that, as the liberation theologians say, "knowing God is doing justice."[6]

NOTES

1. Ananta Kumar Giri, *Knowledge and Human Liberation: Towards Planetary Realization* (London: Anthem Press, 2013). Compare also his *Conversations and Transformations: Toward a New Ethics of Self and Society* (Lanham, MD: Lexington Books, 2002).

2. Commenting on Karl Marx's "Eleventh Thesis on Feuerlach" according to which philosophers in the past have only interpreted the world instead of changing it, Martin Heidegger observed in conversation that "a change of the world presupposes a change of the understanding or conception of the world and that such a conception of the world can only be gained through an adequate interpretation of the world." See Richard Wisser, ed., *Martin Heidegger im Gespräch* (Freiturg-Munich: Alber, 1970), 74.

3. Gustavo Gutierrez, "The Task and Content of Liberation Theology," trans. Judith Condor, in Christopher Rowland, ed., *The Cambridge Comparison to Liberation Theology* (Cambridge, UK: Cambridge University Press, 2007), 29.

4. Clodovis Boff, *Theology and Praxis: Epistemological Foundations* (Portuguese 1978), trans. Robert R. Barr (Maryknoll, NY: Orbis, 1987). See also Zoe Bennett, "'Action is the Life of All': The Praxis-Based Epistemology of Liberation Theology," in Rowland, ed., *The Cambridge Companion to Liberation Theology*, 49.

5. See Felix Wilfred, "Christological Pluralism: Some Reflections," *Concilium*, vol. 3 (2008), 84–94; Trent Schroyer, "Illich's Genealogy of Modern Certitudes," in *Beyond Western Economics* (London: Routledge, 2009).

6. Denys Turner, "Marxism, Liberation Theology and the Way of Negation," in Rowland, ed., *The Cambridge Companion to Liberation Theology*, 246. See also Turner, *The Darkness of God* (Cambridge, UK: Cambridge University Press, 1995).

Appendix C

Theocracy as Temptation: Empire and Mindfulness

> Too long have I dwelled among those who hate peace. I am for peace;
> but when I speak, they are for war. —Psalms 120:6–7

"Theocracy"—meaning God's rule or rulership—is a deeply ambivalent and contested concept. Depending on the meaning assigned to it, the term is liable to generate fierce emotions and antagonisms. To an extent, this outcome may seem surprising. Approached in a calm and dispassionate spirit, the idea of "theocracy" is nothing other than what religious believers—in diverse formulations—routinely affirm. Thus, practicing Christians are wont to recite daily, and perhaps even several times daily, words contained in the "Lord's Prayer" which state: "Your kingdom come, your will be done on earth as it is in heaven." Words or sentiments to a similar effect can be found in several other world religions—without occasioning alarm or disturbance. Here as elsewhere the rub, of course, comes when one tries to define or interpret properly the precise meaning of words. Questions instantly proliferate: What or where is God's kingdom and who is in charge? And even if one grants that God's will should be done in heaven, how can it be done on earth, or who will be the legitimate stand-in for God on earth?

Thus, as one can see, a seemingly routine formula quickly opens up a Pandora's box and even a mine field of queries. Surely, it is proper to tread very cautiously in this field, while searching for reliable guideposts. One such guidepost can be found in the gospel of Matthew in the passage dealing with the temptation of Jesus. Actually, the text mentions several kinds of temptations but culminates in the offer to Jesus of a spectacular type of theocracy. As we read (Matthew 4:8–10), the tempter took him to "a very high mountain" and showed him "all the kingdoms of the world and the glory of them," offering them all to Jesus in return for subservience (a seemingly unequal bargain). We know that the offer was firmly rejected by Jesus with the words: "You shall worship the Lord your God and him alone shall you serve." The passage in Matthew mentions several other tempting offers, such as the promise of food and sustenance—surely an enticing gift after a fasting period of forty days—and probably of an abundance of "consumer items" desirable for commodious living in the world. This offer too, we recall, was rejected with the words that humans "live not by bread alone, but by every word that proceeds from the mouth of God."

All temptations listed by Matthew are memorable, but most memorable and thought-provoking is surely that offer of world-dominion or domination. This offer is particularly thought-provoking because it was not made to just anyone, but to Jesus. Now, for Christian believers, Jesus was not only an exemplary and even a perfect human being—as such he is recognized also by non-Christians—but rather Christ, the "son" of God, or God seen as a human person. It is the same Christ who exhorted his followers to pray that God's "kingdom" may come and God's will be done on heaven and earth. Something curious thus happens in the story of Matthew, for there Christ rejects precisely the opportunity of world-domination which might have seemed the most expedient and promising way to establish God's "kingdom" on earth. Surely, something must dawn on us at this point: namely, that God's kingdom (as invoked in the Lord's Prayer) is something entirely different from world-rule or world-domination. Put differently: God rules entirely rather than worldly potentates, kings, or emperors. This difference was in fact explicitly stated by Jesus in another passage (Matthew 20:25–26): "The rulers of the Gentiles lord it over them, and their great men exercise authority over them. It shall not be so among you."

Biblical stories are sometimes treated as far-off legends, as fables from a by-gone era. The same happens to the passages in Matthew's gospel. The story of the temptations is sometimes viewed as a distant saga, as an exotic event that happened—and could only have happened—to Jesus. Tucked away as a museum piece, it remains without consequences. But how is this possible—as long as Christians are expected not only to pay lip service to, but to follow in the footsteps of their Lord? (And even those who regard him as an exemplary person will surely want to learn from this example.) Thus, seeing that Christ rejected world-domination as a temptation and sinful abomination, why would Christians everywhere not also join in this rejection—no matter by what potentates (kings or emperors) this domination is exercised or attempted to be exercised? How can Christians—or mindful people anywhere—become accomplices in the establishment and expansion of this kind of abomination? Should they not pay heed to the exhortation of Jesus that only the divine (the just and pure) deserves to be worshipped and served? And does this not require of all mindful people a kind of turning or "mind-fasting" as an antidote to the powerful temptation?

To be sure, in our modern world, theocracy comes in many different shapes and forms; not all of them are infested with the same kind of danger or temptation. One way in which theocracy has traditionally manifested itself is in absolutist clerical regimes, that is, in regimes where religious authorities wield absolute or near-absolute political power. Fortunately, under the impact of modern democracy, such regimes are waning (although they have by no means disappeared). In my view, the term should not be applied in situations where religious institutions simply exercise ritual and educational functions in civil society without wielding public control. Unfortunately and distressingly, modern times have generated the near-reversal of traditional (clerical) theocracies: namely, the erection of basically "secular" regimes into quasi-religious, totalitarian structures fueled by comprehensive, sometimes millenarian ideologies. This was the case in the well-known totalitarian systems of the twentieth century. Unhappily, the demise of these systems has not put an end to totalizing global aspirations. The technological "advances" of our time have put at the disposal of political elites unprecedented weapons of mass destruction as well as unheard-of instruments of mass surveillance and mind control. In the case of some elites, these advances have unleashed the dystopia of world-empire, of a total control of the world based on the monopoly of all available knowledge or information as a premise of absolute power. Using the terminology commonly applied to the traditional image of God, one can characterize the dystopia as the linkage of "omniscience" and "omnipotence" — which in turn yields the formula of a "pseudo-theocracy" or sham-theocracy (see chapter 1 above).

What are people placed within the ambit of this imperial dystopia supposed to do? What else, but what Jesus did: dismiss it and move on. A half-way sane or mature person would not wish to waste time in pursuit or in support of worthless affairs. To be sure, there is no point—for mindful individuals—to butt their heads against the ramparts of empire. Given its formidable arsenal of armaments—from nuclear weapons to drones to hidden torture chambers—direct confrontation can only result in self-destruction. At this point, another part of Matthew's story of the temptations of Jesus becomes relevant: where the tempter dared Jesus to throw himself down from the pinnacle of the temple, trusting to be rescued by God's angels (Matthew 4:6–7). This part of the story, in my view, is a warning against willful self-sacrifice or martyrdom, an action which flies in the face of the commandment against killing and self-killing and of the principle of the sacredness of human life. Jesus's clear response to the tempter is that one should not recklessly put God's mercy to the test. If there is not the possibility of direct confrontation—because of its lethal effects—there remains only the possibility of circumspect action, of subterfuge and non-violent resistance, or something which Buddhists traditionally have called "skillful means" (*upaya*).[1]

Given the close connection between world-empire and advanced modern technology, skillfulness here resembles the attitude which Martin Heidegger recommended for the encounter between human beings and the "enframing" power of *Gestell*. Instead of either assaulting that power or pliantly submitting to it, Heidegger counseled a stance of mindful "releasement" or *Gelassenheit*, a stance granting a limited niche to technology in one's life, while otherwise just letting it be or leaving it to its own devices. Left to its own devices and bereft of sustaining enthusiasm, technology's and empire's power might possibly relent and perhaps ultimately subside. Leaving be, one should note, does not in any way signal abandonment of the hope for a different alternative: the hope for something which is the very reverse of pseudo-theocracy or sham-theocracy. This is the hope which animates the search for God's genuine "kingdom," a search which—in the words of St. Augustine—unites people "from all nations and all tongues" into a "harmonious pilgrim band" journeying toward the promised place. Elaborating on Augustine's words, religious philosopher Richard Kearney stresses that the promised abode can neither be engineered or controlled nor be abandoned by human beings: it "can never be fully possessed in the here and now, but always directs us toward an advent still to come."[2] In the here and now, all we can do is keep searching and move on. In this respect, we need to emulate the "wise men from the East" after their visit to Bethlehem: for, knowing about king Herod's cruel designs, "they went to their homes by another way" (Matthew 2:12).

NOTES

1. Compare Michael Pye, *Skillful Means: A Concept of Mahayana Buddhism*, 2nd ed. (New York: Routledge, 2003); John W. Schroeder, *Skillful Means: The Heart of Buddhist Compassion* (Honolulu, HI: University of Hawaii Press, 2001). In many ways, skillful means is similar to the Aristotelian notion of prudent judgment (*phronesis*). One important kind of "*upaya*" was the Gandhi-inspired "non-cooperation movement" with British imperialism in 1920–1922. Another form is the effort to resist and liberate oneself from the pervasive mind-control imposed by empire. International politics expert Hans Köchler speaks in this context of the need to deflate dominant clichés leading to the "deconstruction of imperial myths." See Köchler, "Self-Determination in the Age of Global Empire," http://wpfdc.org/blog/politics/19037-self-determination-in-the-age-of-global-empire.

2. See Richard Kearney, *The God Who May Be. A Hermeneutics of Religion* (Bloomington, IN: Indiana University Press, 2001), 108, 110; St. Augustine, *City of God*, trans. Gerald G. Walsh et al. (Garden City, NY: Image Books, 1958), 465 (Book 21, Chapter 17). Compare also my "Religious Freedom: Preserving the Salt of the Earth," in *In Search of the Good Life: A Pedagogy for Troubled Times* (Lexington, KY: University of Kentucky Press, 2007), 205–219.

Bibliography

Abe, Masao. *Zen and Western Thought*. Edited by William R. LaFleur. Honolulu: University of Hawaii Press, 1985.

Adorno, Theodor W. *Aesthetic Theory*. Translated by Christian Lenhardt. London: Routledge & K. Paul, 1984.

———. *Ästhetische Theorie*. 5th ed. Edited by Gretel Adorno and Rolf Tiedemann. Frankfurt-Main: Suhrkamp, 1981.

———. *Minima Moralia*. Translated by E. F. N. Jephcott. London: New Left Books, 1974.

———. *Negative Dialectics*. Translated by E. B. Ashton. New York: Seabury Press, 1973.

Almond, Gabriel A., and James S. Coleman, eds. *The Politics of the Developing Areas*. Princeton, NJ: Princeton University Press, 1960.

Almond, Gabriel A., and G. Bingham Powell. *Comparative Politics: A Developmental Approach*. Boston: Little Brown, 1966.

Ames, Roger T. *The Art of Rulership: A Study of Ancient Chinese Political Thought*. Albany: State University of New York Press, 1994.

———, ed. *Wandering at Ease in the Zhuangzi*. Albany: State University of New York Press, 1998.

———, and David L. Hall. *Daodejing—"Making This Life Significant": A Philosophical Translation*. Introduced and translated by Roger T. Ames and David L. Hall. New York: Ballantine Books, 2003.

Ananta Murthy, U. K. "Why Not Worship in the Nude? Reflections of a Novelist in His Time." In *Between Tradition and Modernity: India's Search for Identity*, edited by Fred Dallmayr and G. N. Devy, 313–25. New Delhi: Sage, 1998.

The Arabian Nights. Translated by Husain Haddaway. New York: Norton, 1990.

Arendt, Hannah. *Between Past and Future: Six Exercises in Political Thought*. Cleveland: Meridian Books, 1965.

———. *The Human Condition: A Study of Central Dilemmas Facing Modern Man*. Chicago: University of Chicago Press, 1958.

———. *Lectures on Kant's Political Philosophy*. Chicago: University of Chicago Press, 1992.

———. "Truth and Politics." In *The Portable Hannah Arendt*. Edited by Peter Baehr. New York: Penguin Books, 2000, 545-575.

Arkoun, Mohammed. *The Unthought in Contemporary Islamic Thought*. London: Saqi Essentials, 2002.

Augustine. *City of God*. Translated by Gerald G. Walsh et al. Garden City, NY: Image Books, 1958.

Beiner, Ronald. *Political Judgment*. Chicago: University of Chicago Press, 1983.

Benjamin, Walter. "Theses on the Philosophy of History." In *Illuminations*. Edited by Hannah Arendt. Translated by Harry Zohn. 255–66. New York: Harcourt, Brace, 1968.

Bennett, Zoe. "'Action Is the Life of All': The Praxis-Based Epistemology of Liberation Theology." In *The Cambridge Companion to Liberation Theology*, 2nd ed., edited by Christopher Rowland. Cambridge: Cambridge University Press, 2007.

The Bhagavad Gita. Translated by Juan Mascaró. London: Penguin Books, 1962.

Boff, Clodovis. *Theology and Praxis: Epistemological Foundations*. Translated by Robert R. Barr. Maryknoll, NY: Orbis, 1987.

Boyatzis, R. E., and A. McKee. *Resonant Leadership: Renewing Yourself and Connecting with Others Through Mindfulness*. Boston: Harvard Business School Press, 2005.

Buchner, Hartmut, ed. *Japan und Heidegger: Gedenkschrift zum 100. Geburtstag*. Sigmaringen: Thorbecke Verlag, 1989.

The Buddha Speaks. Edited by Anne Bancroft. Boston: Shambala, 2010.

Carroll, M. *The Mindful Leader: Ten Principles for Bringing Out the Best in Ourselves and Others*. Boston: Trumpeter, 2007.

Cézanne, Paul. *Cézanne by Himself: Drawings, Paintings, Writings*. Edited by Richard Kendall. Boston: Little Brown, 1988.

Chuang Tzu. *The Complete Works of Chuang Tzu*. Edited by Burton Watson. New York: Columbia University Press, 1968.

Coyle, Daniel. "On the *Zehnren*." In Ames, *Wandering at Ease in the Zhuangzi*, 197–210.

Dallmayr, Fred. "Action in the Public Realm: Arendt between Past and Future." In Dallmayr, *The Promise of Democracy*, 83–97.

——. "After Babel: Journeying toward Cosmopolis." In Dallmayr, *Being in the World*, 47–58.

——. "Agency and Letting-Be: Heidegger on Primordial Praxis." In Dallmayr, *The Promise of Democracy*, 67–81.

——. "The Ambivalence of Europe: Western Culture and Its 'Other.'" In *Dialogue Among Civilizations: Some Exemplary Voices*. 49–65. New York: Palgrave Macmillan, 2002.

——. "Beautiful Freedom: Schiller on the Aesthetic Education of Humanity." In Dallmayr, *In Search of the Good Life*, 116–37.

——. *Being in the World: Dialogue and Cosmopolis*. Lexington: University of Kentucky Press, 2013.

——. "Beyond Monologue: For a Comparative Political Theory." *Perspectives on Politics* 2 (June 2004): 249–57.

——. *Beyond Orientalism: Essays on Cross-Cultural Encounter*. Albany: State University of New York Press, 1996.

——. "Comparative Political Theory: What Is It Good For?" In *Western Political Thought in Dialogue with Asia*, edited by Takashi Shogimen and Cary J. Nederman, 13–24. Lanham, MD: Lexington Books, 2008.

——. "Democratic Action and Experience: Dewey's 'Holistic' Pragmatism." In Dallmayr, *The Promise of Democracy*, 43–65.

——. "The Future of the Past," *Journal of Contemporary Thought* 12 (Winter 2000): 5–16.

——. "Heidegger and Zen Buddhism." In Dallmayr, *The Other Heidegger*, 200–26.

——. "Heidegger on *Macht* and *Machenschaft*." *Continental Philosophy Review* 34 (2001): 247–67.

——. *In Search of the Good Life: A Pedagogy for Troubled Times*. Lexington: University of Kentucky Press, 2007.

——. *Integral Pluralism*. Lexington: University of Kentucky Press, 2010.

——. "Jacques Derrida's Legacy: 'Democracy to Come.'" In Dallmayr, *The Promise of Democracy*, 117–34.

——. "Multiculturalism and the Good Life: Comments on Bhikhu Parekh." In Dallmayr, *In Search of the Good Life*, 237–45.

——. *The Other Heidegger*. Ithaca, NY: Cornell University Press, 1993.

——. "A Pedagogy of the Heart: Saint Bonaventure's Spiritual Itinerary." In Dallmayr, *In Search of the Good Life*, 23–39.

——. "Political Philosophy Today." In *Polis and Praxis: Exercises in Contemporary Political Theory*. 15–46. Cambridge, MA: MIT Press, 1984.

——. "Postsecular Faith: Toward a Religion of Service." In Dallmayr, *Integral Pluralism*, 67–83.

——. *The Promise of Democracy: Political Agency and Transformation*. Albany: State University of New York Press, 2010.

———. "Religion and the World: The Quest for Justice and Peace." In Dallmayr, *Integral Pluralism*, 85–101.

———. "Religious Freedom: Preserving the Salt of the Earth." In Dallmayr, *In Search of the Good Life*, 205–19.

———. "Resisting Totalizing Uniformity: Martin Heideggar on *Macht* and *Machenschaft*." In *Achieving our World: Toward a Global and Plural Democracy*. 189–209. Lanham, MD: Rowman & Littlefield, 2001.

———. "Speaking Truth to Power: In Memory of Edward Said." In *Small Wonder: Global Power and Its Discontents*. 94–114. Lanham, MD: Rowman & Littlefield, 2005.

———. "The Underside of Modernity: Adorno, Heidegger and Dussel." *Constellations* 11 (2004): 102–20.

———. "Why the Classics Today? Lessons from Gadamer and de Bary." In Dallmayr, *In Search of the Good Life*, 141–53.

Derrida, Jacques. "The Crisis in the Teaching of Philosophy." In *Who's Afraid of Philosophy? Right to Philosophy 1*. Translated by Jan Plug. 99–116. Stanford, CA: Stanford University Press, 2002.

———. "In Praise of Philosophy." In *Eyes of the University: Right to Philosophy 2*, Translated by Jan Plug et al. 157–63. Stanford, CA: Stanford University Press, 2004.

———. *The Other Heading: Reflections on Today's Europe*. Translated by Pascale-Anne Brault and Michael B. Naas. Bloomington: Indiana University Press, 1992.

———. "The Right to Philosophy from a Cosmopolitical Point of View." In *Ethics, Institutions, and the Right to Philosophy*. Edited and translated by Peter Pericles Trifenas. 1–18. Lanham, MD: Rowman & Littlefield, 2002.

———. *Spurs: Nietzsche's Styles/Epérons: Les Styles de Nietzsche*. Translated by Barbara Harlow. Chicago: University of Chicago Press, 1979.

———. "Titles: for the Collège International de Philosophie. 1982." In *Eyes of the University: Right to Philosophy 2*. Translated by Jan Plug et al. 195–215. Stanford, CA: Stanford University Press, 2004.

Desai, Anita. Introduction to Attia Hosain, *Sunlight in a Broken Column*. Delhi: Penguin, 1961.

The Dhammapada. Translated by John Ross Carter and Mahinda Palihawadana. New York: Oxford University Press, 1987.

Die Lehre des Buddha: The Teaching of Buddha. 26th ed. Tokyo: Kosaida Printing, 2006.

Dreyfus, Hubert L. "Heidegger's Ontology of Art." In *A Companion to Heidegger*, edited by H. L. Dreyfus and M. A. Wrathall, 407–19. Oxford: Blackwell, 2005.

Foucault, Michel. *Discipline and Punishment: The Birth of the Prison*. New York: Random House, 1978.

———. "The Eye of Power." In *Power/Knowledge: Selected Interviews and Other Writings 1972–1977*. Edited by Colin Gordon. Translated by Colin Gordon et al. 146–165. New York: Pantheon Books, 1980.

Gadamer, Hans-Georg. "The Marburg Theology." In *Heidegger's Ways*. Translated by John W. Stanley. 29–43. Albany: State University of New York Press, 1994.

———. *Praise of Theory: Speeches and Essays*. Translated by Chris Dawson. New Haven, CT: Yale University Press, 1998.

———. *Truth and Method*. 2nd rev. ed. Translated by Joel Weinsheimer and Donald G. Marshall. New York: Crossroad, 1989.

———. *Wahrheit und Methode*. 2nd ed. Tübingen: Mohr, 1965.

Giri, Ananta Kumar. *Conversations and Transformations: Toward a New Ethics of Self and Society*. Lanham, MD: Lexington Books, 2002.

———. *Knowledge and Human Liberation: Towards Planetary Realization*. London: Anthem Press, 2013.

Graham, A. C. *Chuang-Tzu: The Inner Chapters*. London: Unwin, 1989.

Gutierrez, Gustavo. "The Task and Content of Liberation Theology." Translated by Judith Condor. In *The Cambridge Companion to Liberation Theology*, 2nd ed., edited by Christopher Rowland, 19–38. Cambridge: Cambridge University Press, 2007.

Habermas, Jürgen. *The Past as Future*. Translated by Max Pensky. Lincoln: University of Nebraska Press, 1994.

Halbfass, Wilhelm. "India and the Comparative Method." *Philosophy East & West* 35 (1985): 3–15.

Heidegger, Martin. *Basic Writings*. Edited by David F. Krell. New York: Harper & Row, 1977.

———. *Being and Time*. Translated by John Macquarrie and Edward Robinson. New York: Harper & Row, 1962.

———. *Being and Time*. Translated by Joan Stambaugh. Albany: State University of New York Press, 1996.

———. *Beiträge zur Philosophie. Vom Ereignis*. Edited by Friedrich-Wilhelm von Herrmann. *Gesamtausgabe*, vol. 65. Frankfurt-Main: Klostermann, 1989.

———. *Besinnung*. Edited by Friedrich-Wilhelm von Herrmann. *Gesamtausgabe*, vol. 66. Frankfurt-Main: Klostermann, 1997.

———. *Contributions to Philosophy: of the Event*. Translated by Richard Rojcewicz and Daniela Vallega-Neu. Bloomington: Indiana University Press, 2012.

———. "Das Ding." In *Vorträge und Aufsätze*. 3rd ed. Vol. 2. 37–55. Pfullingen: Neske, 1967.

———. "A Dialogue on Language." In *On the Way to Language*, 1–54. Translated by Peter D. Hertz. New York: Harper & Row, 1971.

———. *Der Begriff der Zeit*. Edited by Hartmut Tietjen. Tübingen: Niemeyer, 1989.

———. "Der Ursprung des Kunstwerkes." In *Holzwege*, 7–9. Frankfurt-Main, 1963.

———. *Die Frage nach dem Ding: Zu Kant's Lehre von den tranzendentalen Grundsätzen*. *Gesamtausgabe*, vol. 44. Frankfurt-Main: Klostermann, 1984.

———. "Die Frage nach der Technik." In *Vorträge und Aufsätze*. Part I. 3rd ed. 5–36. Pfullingen: Neske, 1967.

———. "Die Herkunft der Kunst und die Bestimmung des Denkens." In *Denkerfahrungen*. Edited by Herrmann Heidegger. 147–48. Frankfurt-Main: Klostermann, 1983.

———. "Die Sprache." In *Unterwegs zur Sprache*. 11–33. Pfulingen: Neske, 1959.

———. *Discourse on Thinking*. Translated by John M. Anderson and E. Hans Freund. New York: Harper & Row, 1966.

———. *Einleitung in die Philosophie*. Edited by Otto Saame and Ina Saame-Speidel. *Gesamtausgabe*, vol. 27. Frankfurt-Main: Klostermann, 1996.

———. *Erläuterungen zu Hölderlins Dichtung*. Edited by Friedrich-Wilhelm von Herrmann. *Gesamtausgabe*, vol. 4. Frankfurt-Main: Klostermann, 1981.

———. *Gelassenheit*. Pfullingen: Neske, 1959.

———. *History of the Concept of Time: Prolegomena*. Translated by Theodor Kisiel. Bloomington: Indiana University Press, 1992.

———. "Letter on Humanism." In Heidegger, *Basic Writings*, 241–42.

———. *Mindfulness*. Translated by Parvis Emad and Thomas Kalary. New York: Continuum, 2006.

———. "Modern Natural Science and Technology: Greetings." In *Research in Phenomenology* 7 (1977): 1–4.

———. "On the Essence of Truth." (*Vom Wesen der Wahrheit*). In Heidegger, *Basic Writings*, 117–41.

———. *On Time and Being*. Translated by Joan Stambaugh. New York: Harper & Row, 1972.

———. "The Origin of the Work of Art." In Heidegger, *Basic Writings*, 149–51.

———. *Poetry, Language, Thought*. Translated by Albert Hofstadter. New York: Harper & Row, 1971.

———. "The Question Concerning Technology." In Heidegger, *Basic Writings*, 287–317.

———. *Sein und Zeit*. 11th ed. Tübingen: Niemeyer, 1967.

———. "The Thing." In Heidegger, *Poetry, Language, Thought*, 163–82.

———. *Was heisst Denken?* 3rd ed. Tübingen: Niemeyer, 1971.

————. "Was heisst Denken?" In *Vorträge und Aufsätze*, 3rd ed., Part 2, 3–17. Pfullingen: Neske, 1967.

————. "Was ist Metaphysik?" In *Wegmarken*, 1–19. Frankfurt-Main: Klostermann, 1967.

————. "What Is Metaphysics?" In Heidegger, *Basic Writings*, 97–98.

————. "Wissenschaft und Besinnung." In *Vorträge und Aufsätze*, 3rd ed., Part 1, 37–63. Pfullingen: Neske, 1967.

————. *Zollikoner Seminare: Protokolle-Gespräche-Briefe*. Edited by Medard Boss. Frankfurt-Main: Klostermann, 1988.

Horkheimer, Max, and Theodor Adorno. *Dialectic of Enlightenment*. Translated by John Cumming. New York: Seabury, 1972.

Jochim, Chris. "Just Say No to 'No Self' in *Zhuangzi*." In Ames, *Wandering at Ease in the Zhuangzi*, 35–74.

Kant, Immanuel. "Idea for a Universal History with a Cosmopolitan Purpose." In *Kant's Political Writings*, 41–53.

————. *Kant's Political Writings*. Edited by Hans Reiss. Translated by H. B. Nisbet. Cambridge: Cambridge University Press, 1970.

————. "On the Common Saying: 'This May Be True in Theory, But It Does Not Apply in Practice.'" In *Kant's Political Writings*, 61–92.

Kearney, Richard. *The God Who May Be: A Hermeneutics of Religion*. Bloomington: Indiana University Press, 2001.

————. "Introduction: Ricoeur's Philosophy of Translation." In Paul Ricoeur, *On Translation*. Translated by Eileen Brennan. vii–xx. New York: Routledge, 2006.

Kisiel, Theodore. *The Genesis of Heidegger's Being and Time*. Berkeley: University of California Press, 1993.

Klee, Paul. "On Modern Art." In *Paul Klee: Philosophical Vision: From Nature to Art*. Edited by John Sallis. 9–14. Boston: McMullen Museum of Art, 2012.

Kockelmans, Joseph. *Heidegger on Art and Art Works*. Dordrecht: Nijhoff, 1985.

Lao Tzu. *Tao Te Ching*. Translated by D. C. Lau. New York: Penguin Books, 1963.

————. *Tao Te Ching: A New English Version*. Translated by Stephen Mitchell. New York: Harper Perennial, 1991.

————. *The Way of Life According to Lao Tzu*. Translated by Witter Bynner. New York: Perigee Books, 1972.

Levinas, Emmanuel. *Otherwise Than Being or Beyond Essence*. Translated by Alphonso Lingis. Dordrecht: Kluwer Academic, 1991.

Liu, Jee Loo, and Douglas L. Berger, eds. *Nothingness in Asian Philosophy*. (New York: Routledge, 2014).

Merleau-Ponty, Maurice. "Cézanne's Doubt." In *Sense and Non-Sense*. Translated by Hubert L. and Patricia A. Dreyfus. 9–25. Evanston, IL: Northwestern University Press, 1964.

————. "Eye and Mind." In *The Primacy of Perception and Other Essays*. Edited by James M. Edie. Translated by Carleton Dallery. 159–190. Evanston, IL: Northwestern University Press, 1964.

————. *In Praise of Philosophy*. Translated by John Wild and James M. Edie. Evanston, IL: Northwestern University Press, 1963.

Nietzsche, Friedrich. *Beyond Good and Evil*. Translated by Marianne Cowan. South Bend, IN: Gateway Editions, 1955.

————. "How the 'True World' Finally Became a Fable." In *The Portable Nietzsche*, 485–86.

————. *The Portable Nietzsche*. Edited by Walter Kaufmann. New York: Viking Press, 1968.

————. *Untimely Meditations, Part 2: On the Advantage and Disadvantage of History for Life*. Indianapolis: Hackett, 1980.

Panikkar, Raimon. *The Rhythm of Being: The Gifford Lectures*. Maryknoll, NY: Orbis Books, 2010.

———. "What Is Comparative Philosophy Comparing?" In *Interpreting Across Boundaries: New Essays in Comparative Philosophy*, edited by Gerald J. Larson and Eliot Deutsch, 122–23. Princeton, NJ: Princeton University Press, 1988.

Parekh, Bhikhu. *Rethinking Multiculturalism: Cultural Diversity and Political Theory.* London: Macmillan, 2000.

Parkes, Graham, ed. *Heidegger and Asian Thought.* Honolulu: University of Hawaii Press, 1987.

Petzet, Heinrich Wiegand. *Encounters and Dialogues with Martin Heidegger: 1929–1976.* Translated by Parvis Emad and Kenneth Maly. Chicago: University of Chicago Press, 1993.

Pöggeler, Otto. *Bild und Technik: Heidegger, Klee und die Moderne Kunst.* Munich: Wilhelm Fink Verlag, 2002.

Pye, Lucian W. *Aspects of Political Development.* Boston: Little Brown, 1966.

———, and Sidney Verba, eds. *Political Culture and Political Development.* Princeton, NJ: Princeton University Press, 1965.

Pye, Michael. *Skillful Means: A Concept of Mahayana Buddhism.* 2nd ed. New York: Routledge, 2003.

Ricoeur, Paul. *The Conflict of Interpretations: Essays in Hermeneutics.* Evanston, IL: Northwestern University Press, 1974.

———. *Freud and Philosophy: An Essay on Interpretation.* Translated by Denis Savage. New Haven, CT: Yale University Press, 1970.

Rilke, Rainer Maria. *Letters on Cézanne.* 2nd ed. Edited by Clara Rilke. Translated by Joel Agee. New York: Fromm, 1986.

———. "The Sonnets to Orpheus." In *The Selected Poetry of Rainer Maria Rilke.* Edited and translated by Stephen Mitchell. 234–35. New York: Vintage Books, 1989.

Said, Edward W. *Culture and Imperialism.* New York: Knopf, 1993.

———. *Orientalism.* New York: Vintage Books, 1979.

———. *Representations of the Intellectual: The 1993 Reith Lectures.* New York: Pantheon Books, 1994.

Sallis, John. "Klee's Philosophical Vision." In *Paul Klee: Philosophical Vision*, 17–18.

Schiller, Friedrich. *Don Carlos.* Translated by Frederick W. C. Lieder. New York: Oxford University Press, 1912.

———. *On the Aesthetic Education of Man, in a Series of Letters.* English and German facing. Edited and translated by Elizabeth M. Wilkinson and L.A. Willoughby. Oxford: Clarendon Press, 1967.

Schroeder, John W. *Skillful Means: The Heart of Buddhist Compassion.* Honolulu: University of Hawaii Press, 2001.

Schroyer, Trent. *Beyond Western Economics: Remembering Other Economic Cultures.* London: Routledge, 2009.

Siegel, Ronald D. *The Mindfulness Solution: Everyday Practices for Everyday Problems.* New York: Guilford Press, 2010.

Skaja, Henry G. "How to Interpret Chapter 16 of the *Zhuangzi* 'Repairers of Nature.'" In Ames, *Wandering at Ease in the Zhuangzi*, 102–3.

Stille, Alexander. *The Future of the Past.* New York: Farrar, Straus and Giroux, 2002.

Stirner, Max. *Der Einzige und sein Eigentum* [*The Ego and His Own*]. Edited by John Carroll. Translated by Steven T. Byington. New York: Harper & Row, 1971.

Szondi, Peter. "Hope in the Past: On Walter Benjamin." *Critical Inquiry* 4 (1978): 291–506.

Taylor, Charles. "The Politics of Recognition." In *Multiculturalism and "The Politics of Recognition,"* edited by Amy Gutmann, 25–73. Princeton, NJ: Princeton University Press, 1992.

———. *A Secular Age.* Cambridge, MA: Belknap Press of Harvard University Press, 2007.

Thiele, Leslie Paul. *The Heart of Judgment: Practical Wisdom, Neuroscience, and Narrative.* New York: Cambridge University Press, 2006.

Thomson, Iain D. *Heidegger, Art and Postmodernity.* Cambridge: Cambridge University Press, 2011.

Tillich, Paul. *The Courage to Be.* 2nd ed. New Haven, CT: Yale University Press, 1980.

———. *Systematic Theology. Vol. 1: Reason and Revelation, Being and God.* Chicago: University of Chicago Press, 1951.

Turner, Denys. *The Darkness of God.* Cambridge: Cambridge University Press, 1995.

———. "Marxism, Liberation Theology and the Way of Negation." In *The Cambridge Companion to Liberation Theology,* 2nd ed., edited by Christopher Rowland, 199–217. Cambridge: Cambridge University Press, 2007.

Vetsch, Florian. *Martin Heidegger's Angang der interkulturellen Auseinandersetzung.* Würzburg: Königshausen and Neumann, 1992.

von Herrmann, Friedrich-Wilhelm. *Heidegger's Philosophie der Kunst.* Frankfurt-Main: Klostermann, 1980.

Watson, Stephen M. *Crescent Moon over the Rational: Philosophical Interpretations of Paul Klee.* Stanford, CA: Stanford University Press, 2009.

Wilfred, Felix. "Christological Pluralism: Some Reflections." *Concilium* 3 (2008): 84–94.

Young, Julian. *Heidegger's Philosophy of Art.* Cambridge: Cambridge University Press, 2001.

Index

disclosure, 75–76, 117–119
disinterest, 10, 33
divinization, 10
dread, 44–45

earth, 5, 75–79
egocentrism, 36
Einstein, Albert, 11
emancipation, 123
emptiness, 4, 41, 52–54; of selfhood, 53, 54
enlightenment, 2, 11, 15, 41, 93
epistemology, 14–15, 20, 124–125
equipment, 74–75, 115–116
Ereignis, 46–47, 67–68, 79
essentialism, 53
Eurocentrism, 15
existentialism, 46, 49–50

fatalism, 35
Foucault, Michel, 11
fourfold, 78
Frankfurt School, 93
freedom, 33, 53, 64, 66, 116–119, 123; beautiful, 64
fundamentalism, 5, 35, 83, 87

Gadamer, Hans-Georg, 2, 4, 13–14, 29, 44, 104
Gandhi, Mahatma, 123
Gestell, 6, 21, 118–119, 129
Giri, Ananta, 123–124
globalism, 15
globalization, 5, 83, 87, 94, 99, 104
Goethe, Johann Wolfgang, 63
good life, 102
Gutierrez, Gustavo, 125

Handke, Peter, 63–64
Hegel, Georg F. W., 30, 47, 52, 123
hegemony, 15
Heidegger, Martin, 1, 4–5, 17–24, 30–33, 34, 42–47, 52, 65, 72–83, 103–104, 115–121; *Being and Time*, 17, 42–43, 72, 89, 92; "The Concept of Time", 88–91; *Contributions to Philosophy*, 19, 46; "Dialogue on Language", 88, 95; "Discourse on Thinking", 120–121; "The History of

the Concept of Time", 91; *Introduction to Philosophy*, 17–18, 32–33; "Letter on Humanism", 24, 31; *Mindfulness*, 19–21, 47–48; *On the Way to Language*, 65–69; "The Origin of Art and the Task of Thinking", 78; "The Origin of the Work of Art", 4, 72–77; "Paths Toward Dialogue", 99; "The Question Concerning Technology", 115–120; "Time and Being," 79; *What Calls for Thinking?*, 21–24; "What is Metaphysics?", 43–45; "Zollikon Seminars," 17
Herder, Johann Gottfried, 109
hermeneutics, 12, 49, 90; diatopical, 107; of suspicion, 109
Hinduism, 53
history, 5, 90–91
Horkheimer, Max, 94
humaneness, 18, 109, 110
humanism, 123; apophatic, 125
humanities, 100
humanization, 18, 66, 109
Humboldt, Wilhelm von, 66
Husserl, Edmund, 12, 92

identity, 108; politics of, 108
Illich, Ivan, 125
immanence, 124
imperialism, 104
individualism, 24, 38
interest, 10, 33

justice, 6, 102–103, 125; social, 6

Kant, Immanuel, 11, 15–16, 30, 91, 123–124
karma yoga, 36–39
Kearney, Richard, 107, 129
Kehre, 46–47, 52, 72
Klee, Paul, 5, 72, 79–82
Kleist, Heinrich von, 4, 64
Klopstock, Friedrich Gottlieb, 63
knowledge, 123–124

laissez-faire, 38
language, 4, 62–63, 65–67, 104, 107; poetic, 66
Laozi, 37

About the Author

Fred Dallmayr is Packey J. Dee professor emeritus in the departments of philosophy and political science at the University of Notre Dame. He holds a Doctor of Law degree from the University of Munich and a PhD in political science from Duke University. He has been a visiting professor at Hamburg University in Germany and at the New School of Social Research in New York, and a fellow at Nuffield College in Oxford. During 1991–1992 he was in India on a Fulbright research grant. He is a past president of the Society for Asian and Comparative Philosophy (SACP) and a co-chair of the World Public Forum—Dialogue of Civilizations. His work has been mainly in modern and recent Western philosophy, in comparative philosophy, and in cross-cultural political thought. Among his recent publications are: *Beyond Orientalism* (1996; Indian ed. 2001); *Achieving Our World* (2001); *Dialogue among Civilizations* (2002); *In Search of the Good Life* (2007); *The Promise of Democracy* (2010); *Return to Nature* (2011); and *Being in the World: Dialogue and Cosmopolis* (2013).

CPSIA information can be obtained at www.ICGtesting.com
Printed in the USA
BVOW01*0817120914

366245BV00007B/4/P

9 780739 199862